Humanizing Artificial Touch

Skin, Sensing, and Structures

Humanizing Artificial Touch

Skin, Sensing, and Structures

John-John Cabibihan
Department of Mechanical and Industrial Engineering
Qatar University
Doha, Qatar

ELSEVIER
ACADEMIC PRESS
An imprint of Elsevier

Academic Press is an imprint of Elsevier
125 London Wall, London EC2Y 5AS, United Kingdom
50 Hampshire Street, 5th Floor, Cambridge, MA 02139, United States

ISBN: 978-0-323-95519-5

For information on all Academic Press publications
visit our website at https://www.elsevier.com/books-and-journals

Publisher: Mara Conner
Acquisitions Editor: Sonnini Yura
Editorial Project Manager: Sudip Saha
Production Project Manager: Bharatwaj Varatharajan
Cover Designer: Greg Harris

Typeset by VTeX

To the ones who matter most:

Melody, Leo Angelo, Marie Françoise and David Lee.

Cover and part illustrations

When I was a child growing up in the 1980s, *Star Wars*, *The Empire Strikes Back* made a strong impression on me. In their final duel, Luke Skywalker and Darth Vader battled on a bridge with their lightsabers. The contrast was stark: Luke had a blue-lit saber against Vader's red. Darth Vader swung his lightsaber, hitting Luke and severing his right arm.

The movie's final scenes showed Luke's arm being reconstructed. A medical droid performed tests on Luke's newly connected prosthetic arm. The robot poked a sharp probe on Luke's fingertips and palm; the fingers twitched; the reflex system was working. Luke stood up, went to his sister, Princess Leia. Luke placed his prosthetic arm on Leia's shoulder. Decades later, I found myself in the intersection of Biorobotics and Social Robotics.

The ideas for this book's front cover took time to develop. Revisiting my old notes and presentations, I rediscovered the artwork on the front page of my PhD defense slides in Sant'Anna Pisa, in May 2007 and my presentation for Assistant Professorship application at the National University of Singapore in October of that year. Way back then, I liked the picture because I imagined it to be a Jedi's hands—Luke Skywalker's hands.

The original artwork, Study of Hands (pen and ink on paper), was drawn in 1506 by the Renaissance artist Albrecht Dürer, who was born in Nuremberg, Germany, in 1471 and lived until 1528. Dürer is renowned for his hand studies, making his work a fitting choice for the book's cover.

I selected some of his sketches to be shown as the cover figure for each Part of this book. For the Introduction and Conclusion, I chose Dürer's studies of hands that are holding open and closed books (1506). For Part II—Skin, I selected Dürer's Arm of Eve (1507). For Part III—Sensors and for Part IV—Structures, I selected Dürer's study of hands of the Father for Heller's altarpiece (1508).

Those figures were also selected to depict how artificial hands or fingers have been used to restore the functionality of missing appendages in recent years. Some of the selected drawings also depict my imaginings of how artificial hands and arms for prosthetics or humanoid robotics can be humanized in the near future.

To illustrate the "humanizing artificial touch" aspects, I persuaded an up-and-coming artist, Marie Françoise Cabibihan, to adapt the figures for the theme of this book. I cautioned her to do her best on the adaptations to give honor to the original works of Albrecht Dürer. All of Dürer's works that were used and adapted in this book were obtained from the public domain. Image source: WikiArt (https://www.wikiart.org/).

Contents

Part III
Sensing

Biography

John-John Cabibihan

John-John Cabibihan is a Professor in the Department of Mechanical and Industrial Engineering at Qatar University. He received the Ph.D. degree in Bioengineering, with specialization in Biorobotics, from the Scuola Superiore Sant'Anna, Pisa, Italy, in 2007. During his Ph.D. studies, he was admitted to the cotutelle Ph.D. programme of École Normale Supérieure Paris–Saclay, France, in 2004–2005, where he conducted research on soft materials. In 2021, he completed the Venture Creation and Finance Programmes at the Saïd Business School, University of Oxford, United Kingdom.

Prior to joining Qatar University, he served as an Assistant Professor in the Department of Electrical and Computer Engineering at the National University of Singapore and as Deputy Director of the Social Robotics Lab. Dr. Cabibihan has secured more than US$8.3M in competitive research funding and has authored or coauthored over 200 peer-reviewed publications. He is an inventor in 20 patent families and has edited or coedited more than 15 books. He was Assistant Editor and Editorial Board Member of the International Journal of Social Robotics and has played a leading role in the growth of the International Conference on Social Robotics community since its inception in 2009. He has served as General Chair, Program Chair, or Finance Chair of major robotics, automation, and control conferences across Asia, Europe, and North America. In 2025, Dr. Cabibihan was appointed Vice President for Finance of the Asian Control Association.

His contributions to robotics and human-centered technologies have been featured in Nature Materials, MIT Technology Review, British Broadcasting Corporation (BBC), Popular Science, New Scientist, Discovery News, and New York Magazine. His research spans social robotics, human-robot interaction, tactile intelligence, embodied artificial intelligence, and additive manufacturing. He has contributed to the development of artificial touch technologies, prosthetic hands, and socially interactive robots designed to enhance quality of life and promote human dignity.

Preface

A grand challenge in prosthetics and social robotics is to create artificial hands that can establish an emotional connection with humans. To achieve a more natural and rich human-robot tactile interaction, such artificial hands need humanlike features. Humanizing Artificial Touch: Skin, Sensing, and Structures presents the advances in artificial skin modeling, robotic tactile sensing technologies, and their integration into the structure of an artificial hand mimicking the human hand's important features. This book gives a special treatment to the relationship between the artificial skin material and the embedded tactile sensors. Furthermore, the book describes how the artificial skin and the embedded sensors can be designed and constructed using 3D printing technologies with case examples on how these were created.

This book is intended to be a useful introductory reference for graduate and postgraduate students, engineering and robotics students, and research scholars who are embarking on the multidisciplinary fields of bioengineering, rehabilitation engineering, and humanoid robotics. This book will also be a valuable reference for those who are doing research on artificial skins, tactile intelligence, robot hand design, and embodied artificial intelligence.

There are five parts to this book. Part I—Introduction lays down the rationale for why humanizing artificial touch can provide useful insights for prosthetic and humanoid robotic hands. It also outlines the book's bottom-up architecture across Skin, Sensing, and Structures. Part II—Skin is concerned with the biomechanical inspiration, mechanical modeling of the skin materials, and the properties of skin conformance, compliance, and hysteresis. Moreover, the time-dependent behavior and the effect of the artificial skin's thickness will be presented. Part III—Sensing presents artificial tactile sensing beginning with the inspiration source, the human mechanoreceptors. Tactile sensing will be treated according to the sensing of the finger-object and object-environment interactions, object slippage, and sensing of the local and global object shape primitives. Part IV—Structures discusses some of the methodologies to create artificial hand structures. It begins with a review of human and artificial hands and then proceeds with the construction techniques for finger phalanges, finger and hand structures, and artificial hands with soft and warm characteristics. Part V—Conclusions synthesizes the foundational mimicry concepts and the prerequisite technologies for the development of bio-inspired prosthetic and robotic hands. This part also describes the static and dynamic Humanizing Artificial Touch (HAT) Tests as a protocol to evaluate artificial touch that can be considered indistinguishable from human touch.

It is with extreme gratitude that I acknowledge all my mentors, colleagues, collaborators, research students, and research funding offices for their continued support and collaboration. I am grateful to the team at Elsevier for their editorial support in making this book project a reality.

John-John Cabibihan
Doha, Qatar
March 2026

Part I

Introduction

Chapter 1

Introduction

1.1 Human touch in action

How do we start our day?

From the moment we awaken, our hands are in constant movement. Within the first 10 minutes of the day, our hands engage in a flurry of tasks, from rubbing our eyes and pushing ourselves out of bed to the fine motor control needed for personal hygiene, like using a faucet, the toothpaste and toothbrush, and brushing our teeth. This initial burst of activities with our hands plays an essential role in our everyday lives. Research has identified at least 20 grasps used for the typical activities of daily living [1][2][3].

The extensive use of our hands highlights the intricate sensorimotor capabilities and complex coordination required for even the simplest daily tasks. Let's take the example of drinking coffee from a cup (Fig. 1.1). This activity involves several coordinated steps.

First, the eyes set gaze targets on objects before kinematic events occur for grasping [4]. The target location depends on the object's affordances. For the coffee cup, which is presumably hot, the target location for grasping would be at the protruding ear-shaped handle. The eyes send visual information to the brain to identify and localize the object in space [5]. The visual cortex processes this information and sends it to the motor cortex. The brain sets the target goals for grasping the cup's handle. The forces to be applied are estimated based on the approximate weight of the cup and the volume of coffee inside the cup. This step involves the premotor cortex and supplementary motor area to plan the subsequent movements [6].

FIGURE 1.1 Human hand reaching and preshaping for a coffee cup.

Humanizing Artificial Touch. https://doi.org/10.1016/B978-0-32-395519-5.00010-6

The brain then integrates the visual information with the sensory feedback from the hand for the appropriate coordination of the eyes and the hand [7]. The motor cortex sends the command signals through the corticospinal tract to the spinal cord and then to the muscles of the arm and hand [8]. The arm reaches for the coffee cup through the coordinated movements of the shoulder, elbow, and the wrist joints. This involves the activation of the deltoid, biceps, triceps, and the forearm muscles [9].

As the hand approaches the cup's handle, the fingers begin to orient themselves in accordance with the shape of the handle of the cup (i.e., preshaping). The angle of approach also varies depending on the height of the handle relative to whether the person is sitting or standing, on the approximate weight of the cup, and on the volume of the cup's content. The hand's preshaping behavior is based on the visual information and prior knowledge of the object's size, shape, and material properties [10].

Upon contact with the object, the mechanoreceptors provide feedback about the cup handle's texture, curvature, edges, and surface moisture [11]. This feedback from the surface contact helps modulate the grip force to prevent the cup from tilting or slipping.

Compared to drinking from a cylindrical glass, a coffee cup causes a downward rotation at the point of contact because the center of mass of the cup and the coffee is away from the fingers. The brain continuously adjusts the forces applied by the fingers and thumb to lift and stabilize the cup [12]. As the cup is lifted from the table's surface to the lips, dynamic force adjustment is needed to ensure that the cup is stabilized and the hot coffee or tea is not splashed onto the face or spilled on clothes.

Johansson [13] proposed a model for the predictive feedforward sensory control of manipulation (Fig. 1.2) to further explain the control mechanisms in tasks such as our coffee cup example. This model describes the mechanism involved in stabilizing the grasp of objects. Two pathways were proposed for this process. The feedforward path was given the term anticipatory parameter control (APC) and the feedback path was given the term discrete event, sensory-driven control (DESC).

When the target object is seen, the human vision system activates the automatic retrieval of the relevant internal models to parametrically adapt the treatment of the object through the motor commands. Seeing the object activates the selection of general action programs, based on the initial assumption regarding the physical properties of the object. These two elements provide information for the motor programs to anticipate the required forces to perform the task. Thus, the feedforward path was termed anticipatory parameter control.

Once the arm reaches for the object and the fingers touch the target locations on the object, the internal models related to the object properties get updated due to the signals detected by the tactile sensors in the hand. This feedback process involves a comparison between the somatosensory information inflow and the predicted inputs. The mismatch between the predicted and actual sensory input

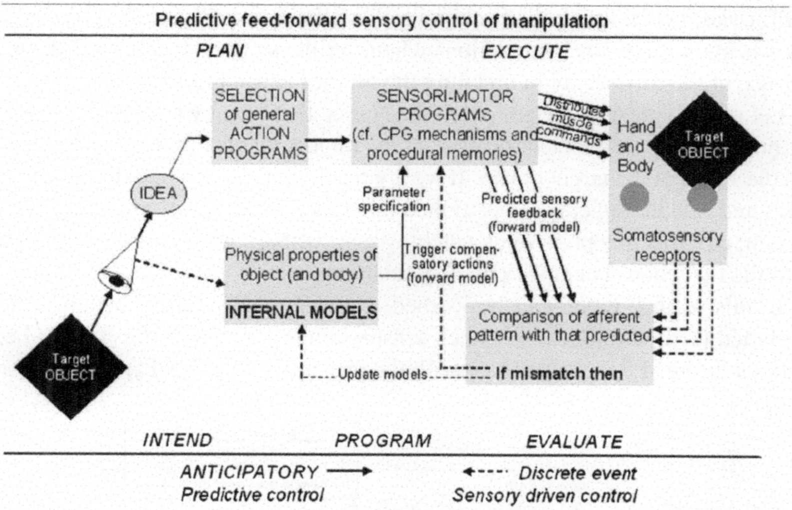

FIGURE 1.2 Predictive feedforward sensory control of manipulation [13]. The sight of a target object activates two paths through the selection of general action programs and an assumption of the physical properties of the object. Both of these initial assumptions proceed to the predictive feedforward sensory motor programs that send commands to the muscles, which would grasp and/or manipulate the target object. The hand, with its rich and responsive tactile sensors, proceeds to a process of comparing the predicted and the actual somatosensory information from the contact with the target object. A mismatch between the predicted and actual inputs results in two paths. The first triggers compensatory action (e.g., slippage control). The second path updates the internal models based on the object's actual properties (e.g., weight) instead of the initially predicted ones. Reproduced from Laschi et al. [14], with permission.

results in two paths. The first path is to trigger the compensatory and corrective actions for the sensorimotor programs to make the adjustments to the muscle commands. The second path goes back to update the internal models of the object's physical properties based on the contact information from the tactile sensors. The feedback path was termed discrete event, sensory-driven control.

1.2 Humanizing artificial touch

The human hand is the foremost interface of our bodies to the world. With our hands, we can both cuddle a baby and hammer a nail on the wall. The artificial hands for humanoid robots and for prostheses are expected to touch and be touched by humans. Those hands are expected to work effectively on objects in human-centered environments.

In the design of artificial hands, the human hand offers the best initial model. Its design and performance have been optimized for dexterity in object grasping and manipulation, fast adaptability to sudden changes during contact, and it conveys softness and warmth in social interactions.

Biological systems are exceptionally remarkable in dealing with complex and dynamic situations and environments. They implement both predictive feedforward control and feedback control mechanisms. The predictive control relies heavily on sensory predictions whenever the sensory feedback is delayed because of the time-delay in updating the motor control [15][13].

Let us take preshaping, for example. Notice that when we reach for objects, our fingers have preselected the pose of the fingers that is optimized for the size and the shape of the target object. If we are reaching for a pen, the thumb and the index and middle fingers are positioned in a pinch grasp to get a stable grasp of the pen; contrast this pose when catching a basketball, where all the ten fingers from the two hands become open and outstretched.

In this regard, much can be learned from the neuroscience of touch. The combined feedforward and feedback sensory control model in Fig. 1.2 has been redrawn in Fig. 1.3 in accordance with the principles of control systems theory.

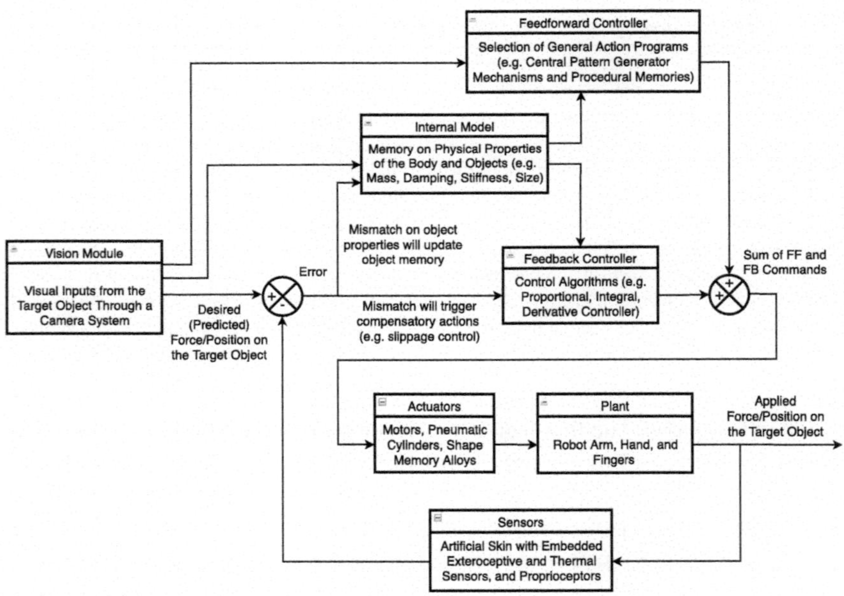

FIGURE 1.3 Integrated feedforward–feedback control architecture for visually guided robotic grasping. Redrawn from the predictive feedforward sensory control model as proposed by Johansson [13], this version includes a vision module that generates signals that get passed to the feedforward controller and the block for the internal model that stores information about the object. Both the feedforward controller and the system's memory of the object generate the predicted force and/or position signal to actuate the robot hand. For any mismatch between the reference predicted signal and the actual signal, the feedback controller (e.g., PID) modulates the error signal by performing corrective responses, such as slippage compensation. The mismatch also updates the object memory to improve future actions.

Instead of human eyes, a camera-based system with real-time video processing (i.e., vision module) provides information about the target object. The information is then passed to an internal model block that stores the parameters related to the physical properties of the object (e.g., mass, damping, stiffness, geometry). Additionally, information is passed to the feedforward controller that selects motion programs to produce an anticipatory action to grasp or manipulate the target object through the robotic hand (e.g., reaching and preshaping algorithms).

The contact information from the robotic hand and the object is detected by the sensors. The sensors, together with the skin covering, are responsible for determining the contact information related to the magnitudes and directions of contact forces, which are consequential to the stable grasping and control of an object. This sensor block also detects the positions and angles of the joints as the hand grasps an object. The information from the feedback path goes to a summing junction that compares the reference signal from the desired force and/or position of the target object, $r(t)$, and the actual force and/or position of the target object, $y(t)$. The difference results in the tracking error, $e(t) = r(t) - y(t)$.

That mechanism is analogous to the comparator mentioned in the neuroscientific model in Fig. 1.2 to compare mismatches in the signals. When the object starts to slip, the mismatch triggers a signal to the feedback controller to implement compensatory actions to the motors of the robot hand to adjust the angles on the motors to increase the grasp force so the object will not fall.

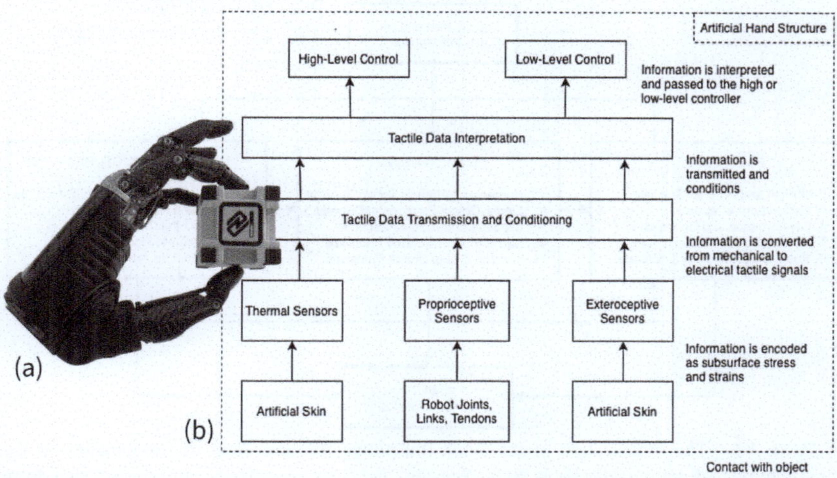

FIGURE 1.4 A robot hand in contact with an object and the schematic flow of information on that contact interaction. (a) The fingers of a robotic hand in physical contact with the target object. (b) The bottom-up flow of information from the skin, sensors, and the structure of the robot hand.

The approach taken in this book begins by considering the information flow when the artificial hand gets into contact with an object (Fig. 1.4a). Referring to the control system diagram (Fig. 1.3), the coverage of this book is in the plant, the sensors, and the controller. When the artificial skin touches the object, the mechanical signals get transmitted to the embedded tactile sensors. The general sequence of the transmission of information is shown in Fig. 1.4b. Exteroceptive sensors (exteroceptors) detect the interactions between the hand, object, and the environment. These sensors are embedded on the skin of the robot fingers. Proprioceptive sensors (proprioceptors) are the sensors for detecting joint angles and torques. Thermal sensors are sensors for detecting the temperatures at the contact surface with the object.

From the bottom to the top, the skin comes into contact with the object. This leads to a distribution of forces on the skin. The surface pressure or deflection results in stresses and strains at the subsurface of the skin. Then, the tactile sensors detect the mechanical signals and convert those signals into electrical signals. The practicalities of signal transmission and conditioning are dealt with through various techniques for filtering, multiplexing, and wiring. These are then processed and are sent for further implementation on the high-level or low-level controllers of the hand. From Fig. 1.3, the high-level controller can represent the path following the internal model (e.g., recognizing an object's shape) to the feedforward controller. The low-level controller can represent the feedback controller for mitigating the object's slippage, for example.

1.3 Book outline

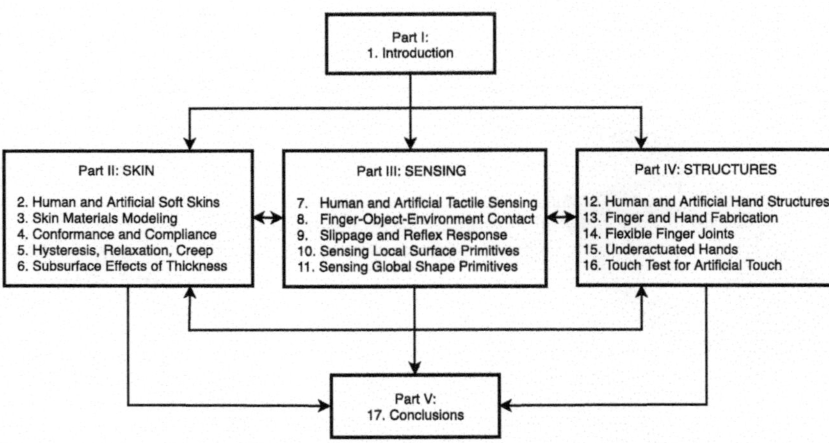

FIGURE 1.5 The organization of this book illustrating the flow from the introduction to the three interrelated parts of the book: Skin, Sensing, and Structures. Each part covers topics from the mechanics of soft skin materials, to tactile sensing, and to the fabrication of finger and hand structures. The bidirectional flow emphasizes the interdependencies among the skin, sensors, and the physical structures, which then converge to the conclusions chapter that gives the concluding remarks, recommendations, and the future outlook.

Fig. 1.5 shows the general framework of the organization of this book. There are three main parts to this book: Skin, Sensing, and Structures. There is an interrelationship on how each part contributes to the other. As shown in the information flow in Fig. 1.4, there are relationships between the skin and sensors, between the sensors and the structure, and between the skin and the structure. The interested reader can proceed to a combination of parts and/or chapters as needed. However, each part or chapter can be treated as a stand-alone reading. For example, those who are interested in hand and finger fabrication can proceed to Chap. 13. Those interested in the finite element analysis on the artificial skins can proceed to Part II. The introductory chapters of each part give an overview of the human and artificial skin, sensing, and structures. It would be a good idea to start with those introductory chapters to get an overview of the succeeding four chapters in each of the main parts.

Part II, Skin, opens this book through a discussion of biological and soft engineering materials. Chap. 2 presents the human skin and its layered anatomy, functionality, and microfeatures, like the fingerprints. The chapter then reviews its closest equivalent materials, which are silicone and polyurethane rubbers. Chap. 3 presents the techniques used for constitutive modeling, finite element simulations, and validation experiments of those models. Chap. 4 talks about the human skin's conformance and compliance properties and how we can use those properties to find the artificial equivalent. Chap. 5 discusses the time-dependent properties of the human fingertip skin. These are hysteresis, stress relaxation, creep and recovery. Likewise, we check whether these properties are present in the artificial skin samples that we experimented on. Chap. 6 completes the skin part by examining the effect of the skin thickness in understanding the subsurface consequences from the contact at the skin surface.

Part III, Sensing, shifts the discussions from the mechanics and simulations of soft skin materials to tactile intelligence. Chap. 7 introduces the types of mechanoreceptors on the non-hairy part of the human hand (i.e., at the palm side). Various types of artificial tactile sensors are then described. Chap. 8 discusses contact sensing at two levels: the contact between the finger and object, and the contact between the object and the table. Chap. 9 presents the slippage mechanisms of the human hand. The artificial slippage sensing and the algorithms for reflex response are introduced. Chap. 10 presents the discrimination of local shape through fingerprint-like ridges. Chap. 11 presents the recognition of various global shapes through sensorized gloves that can approximate the sensing capabilities of proprioceptive sensors of the human hands.

Part IV, Structures, integrates skin and sensing into a complete hand embodiment. Chap. 12 reviews the musculoskeletal system of the human hand (i.e., bones, cartilage, ligaments, tendons and muscles) and their analogous components for the structures of the artificial hand (i.e., motors, tendon-cable systems, etc.). Chap. 13 describes the fabrication techniques to construct fingers and hands using scanning approaches and additive manufacturing technologies. Chap. 14 presents the design and analyses of flexible joints with a case study on the flexible joints during a natural pose. Chap. 15 presents an analysis of underactuation of artificial hands and the possible grasp poses that can be made for the openly accessible 3D printed hands. Chap. 16 concludes the structures part with a perceptual evaluation of lifelike touch with thermo-soft artificial hands, which have been shown to be indistinguishable from a human's touch.

This book concludes with Part V, Conclusions. Chap. 17 presents the concluding remarks, summarizes the key contributions, discusses the limitations of the work, and outlines directions for future research. The chapter also presents guidance for conducting the Humanizing Artificial Touch (HAT) Tests. The book closes with an Epilogue that invites readers to reflect on the broader implications of humanized artificial touch and the questions it raises for the future of technology, human–robot interaction, and society.

References

[1] R.C. Lyle, A performance test for assessment of upper limb function in physical rehabilitation treatment and research, International Journal of Rehabilitation Research 4 (1981) 483–492, https://doi.org/10.1097/00004356-198112000-00001.

[2] C. Sollerman, A. Ejeskär, Sollerman hand function test: a standardised method and its use in tetraplegic patients, Scandinavian Journal of Plastic and Reconstructive Surgery and Hand Surgery 29 (1995) 167–176, https://doi.org/10.3109/02844319509034334.

[3] C.M. Light, P.H. Chappell, P.J. Kyberd, Establishing a standardized clinical assessment tool of pathologic and prosthetic hand function: normative data, reliability, and validity, Archives of Physical Medicine and Rehabilitation 83 (2002) 776–783, https://doi.org/10.1053/apmr.2002.32737.

[4] R.S. Johansson, G. Westling, A. Bäckström, J.R. Flanagan, Eye–hand coordination in object manipulation, The Journal of Neuroscience 21 (2001) 6917–6932, https://doi.org/10.1523/JNEUROSCI.21-17-06917.2001.

[5] M.A. Goodale, A.D. Milner, Separate visual pathways for perception and action, Trends in Neurosciences 15 (1992) 20–25, https://doi.org/10.1016/0166-2236(92)90344-8.

[6] G. Rizzolatti, G. Luppino, The cortical motor system, Neuron 31 (2001) 889–901, https://doi.org/10.1016/S0896-6273(01)00423-8.

[7] R.A. Andersen, Cui H. Intention, Action planning, and decision making in parietal–frontal circuits, Neuron 63 (2009) 568–583, https://doi.org/10.1016/j.neuron.2009.08.028.

[8] R.N. Lemon, Descending pathways in motor control, Annual Review of Neuroscience 31 (2008) 195–218, https://doi.org/10.1146/annurev.neuro.31.060407.125547.

[9] M. Jeannerod, The timing of natural prehension movements, Journal of Motor Behavior 16 (1984) 235–254, https://doi.org/10.1080/00222895.1984.10735319.

[10] P. Jenmalm, R.S. Johansson, Visual and somatosensory information about object shape control manipulative fingertip forces, The Journal of Neuroscience 17 (1997 Jun) 4486–4499, https://doi.org/10.1523/jneurosci.17-11-04486.1997.

[11] R.S. Johansson, J.R. Flanagan, Coding and use of tactile signals from the fingertips in object manipulation tasks, Nature Reviews. Neuroscience 10 (2009 Apr) 345–359, https://doi.org/10.1038/nrn2621.

[12] G. Westling, R.S. Johansson, Responses in glabrous skin mechanoreceptors during precision grip in humans, Experimental Brain Research 66 (1987) 128–140, https://doi.org/10.1007/BF00236209.

[13] R.S. Johansson, Sensory Input and Control of Grip, Wiley, 2007, pp. 45–63, https://doi.org/10.1002/9780470515563.ch4.

[14] C. Laschi, G. Asuni, E. Guglielmelli, G. Teti, R. Johansson, H. Konosu, Z. Wasik, M.C. Carrozza, P. Dario, A bio-inspired predictive sensory-motor coordination scheme for robot reaching and preshaping, Autonomous Robots 25 (2007) 85–101, https://doi.org/10.1007/s10514-007-9065-4.

[15] A. Berthoz, Le sens du mouvement, 2008, Editions Odile Jacob.

Part II

Skin

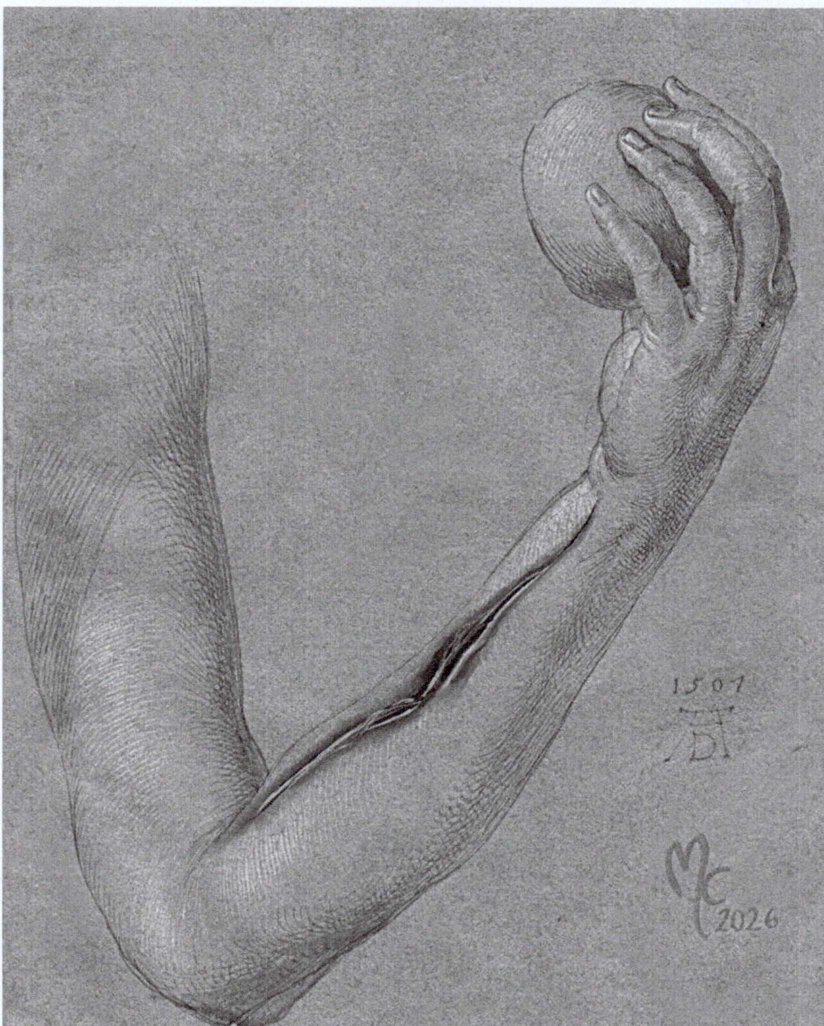

Chapter 2

Human and artificial soft skins

2.1 Introduction

The human skin is the first part of the body that makes physical contact with the external environment, with objects, and even with another human. The human skin is the largest organ of the body. It has multiple purposes. Functionally, the skin acts as protection, provides sensation, thermoregulation, and biochemical synthesis [1]. To support these functional roles, the skin has a complex composition consisting of a multilayered structure: the epidermis, dermis, and subcutaneous layer [2][3].

The outermost layer, the epidermis, is composed of keratinocytes [4]. The epidermis serves as the barrier against external threats while undergoing periodic regeneration and exhibiting self-healing properties when punctured or lacerated. Beneath it is the dermis layer, which contains a network of collagen and elastin fibers [5]. This layer provides structural support and elasticity. Beneath the dermis layer is the subcutaneous layer or the hypodermis. This layer connects the upper layers of the skin to the underlying tissues [3]. The subcutaneous layer provides cushioning and thermoregulation.

The human skin's ability to absorb and dissipate energy can be attributed to the skin's viscoelastic behavior. When implemented in artificial skins, the skin material will similarly absorb and dissipate the energy coming from contact with objects and the external environment. Moreover, the viscoelastic behavior has an effect on the embedded tactile sensors. Typically, mechanical models, such as Kelvin-Voigt and generalized Maxwell models, have been used to describe the viscoelastic behavior of the skin [3].

The characteristics of the human skin offer many potential benefits for the artificial skins of robots or prosthetic arms. To begin with, a basic understanding of the mechanical properties of the human skin can help in the selection of suitable soft skin materials. A further investigation can be carried out to determine whether the selected materials are acceptable for embedding tactile sensors and for covering the hand structure.

This chapter has been divided into two parts: human skin (Sec. 2.2) and artificial skins (Sec. 2.3). The useful characteristics for both are described in the subsections.

Humanizing Artificial Touch. https://doi.org/10.1016/B978-0-32-395519-5.00012-X

2.2 Human skins

2.2.1 Layer organization

The structure of the human skin is a highly integrated and layered structure. Fig. 2.1 shows the schematic organization of the layers of the two types of human skin. Both types have structural and functional similarities [6]. Both the hairless and hairy types consist of the same primary layers: epidermis, dermis, and subcutaneous tissue. Likewise, both types of layers serve as barriers to environmental elements, pathogens, and loss of water. Moreover, both types of skin contain sensory receptors that are responsible for the human sense of touch.

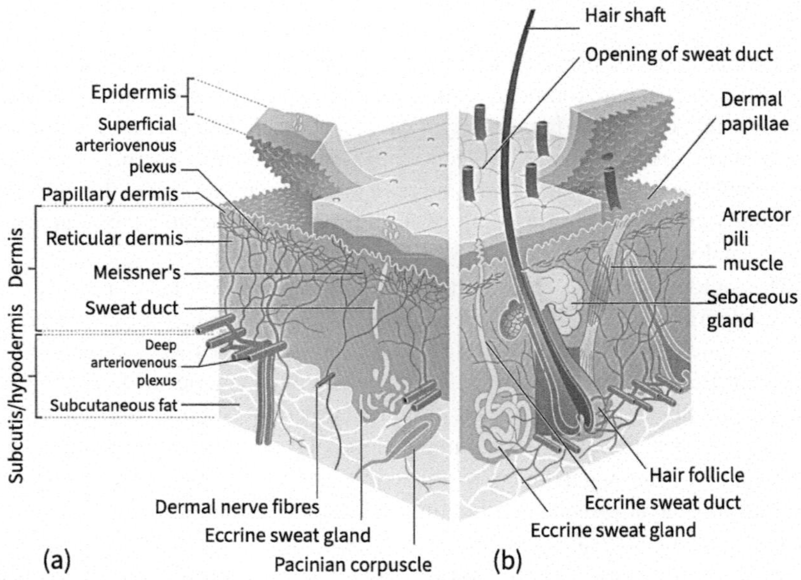

FIGURE 2.1 Schematic representation of the glabrous (hairless) and hairy skin. (a) Illustration of the glabrous skin, which is found on the palms of the hand and soles of the feet. (b) Illustration of the hairy skin. Adapted from "Skin layers" by Madhero88, Wikimedia Commons, licensed under CC BY-SA 3.0.

2.2.1.1 Glabrous (hairless) skin

The hairless or glabrous type of the human skin can be found on the palms of the hands and on the soles of the feet. The glabrous skin is thicker and has tougher mechanical properties to adapt to the contact interactions with the external environment [6]. Starting from the surface of the palm (Fig. 2.1a), the layers include the stratum corneum, epidermis, dermis, and a layer of fat. The epidermis has a high concentration of mechanoreceptors (i.e., Merkel cells, Meissner corpuscles, Ruffini endings, and Pacinian corpuscles), which are known to enhance the tactile sensitivity for grasping and manipulation [7].

2.2.1.2 *Hairy skin*

The hairy type of skin can be found on most parts of the body (Fig. 2.1b). This type contains hair follicles, which are related to sebaceous glands that secrete sebum [6]. Compared to the hairless skin type, the hairy type is generally thinner and has a more dispersed distribution of mechanoreceptors. The hair follicles contribute to the overall appearance and texture of the hairy skin. The associated glands contribute to keeping the skin hydrated and protected. The hairy skin also has a functional role for thermoregulation and affective tactile perception. The C-tactile (CT) afferents play an important role in the perception of gentle, affective touch, which is associated with positive emotions [8] [9].

2.2.2 Skin layers and characteristics of the glabrous skin

2.2.2.1 *Stratum corneum*

The stratum corneum is a rough, protective layer of dead skin cells, which are continually discarded by the body. Mechanically, the stratum corneum exhibits viscoelastic behavior. The stress-strain relations of this layer were shown to be rate-dependent (i.e., it becomes stiffer with increasing strain rate) [10]. The overall contribution of the thin stratum corneum is small compared with the thick dermis layer [11]. Harkness et al. [11] also observed that the stratum corneum tears prior to the dermis layer in rupture tests.

2.2.2.2 *Epidermis layer*

The epidermis is the dominant layer in determining frictional properties, which can change significantly under different moisture conditions due to its ability to absorb a considerable amount of water [12]. It was shown that the friction coefficient increases due to sweat, moderate hydration, and the removal of grease, but decreases with wetting, dry talc powder and the addition of oil [13]. Despite its smaller thickness, the epidermis is comparatively stronger than the dermis [14].

2.2.2.3 *Dermis layer*

Lanir [1] describes the dermis of the glabrous skin as containing collagen, elastin, reticulin, blood vessels, nerve endings, and sweat glands. Collagen has been observed to be an apparently disordered three-dimensional network of coiled fibers. Upon stretching the skin, the response of the collagen fibers predominates and accordingly these fibers are considered the major mechanical elements of the skin. They are strong, stiff and can stretch reversibly up to 2 to 4%. They are viscoelastic in their native state [15][16]. While the collagen fibers are still unstretched at low levels of strain, it is believed that the elastin components are the first to be stretched when the skin tissue is strained. The elastic properties of elastin are similar to those of rubber and rubberlike materials; it is less stiff than collagen and can be reversibly stretched by more than 100% [17].

2.2.2.4 Subcutaneous tissue

The subcutaneous tissue is composed of fat cells, which form a cushioning layer between the skin and the deeper muscles. The subcutaneous tissue was estimated to contain 60-72% fluid in volume [18]. It is believed that fluidlike substances diffuse from the interfiber space while the fibers become reoriented and densely packed when the tissue is stretched [1]. The subcutaneous tissue was modeled to be a sponge-like porous material that is saturated with fluid [19].

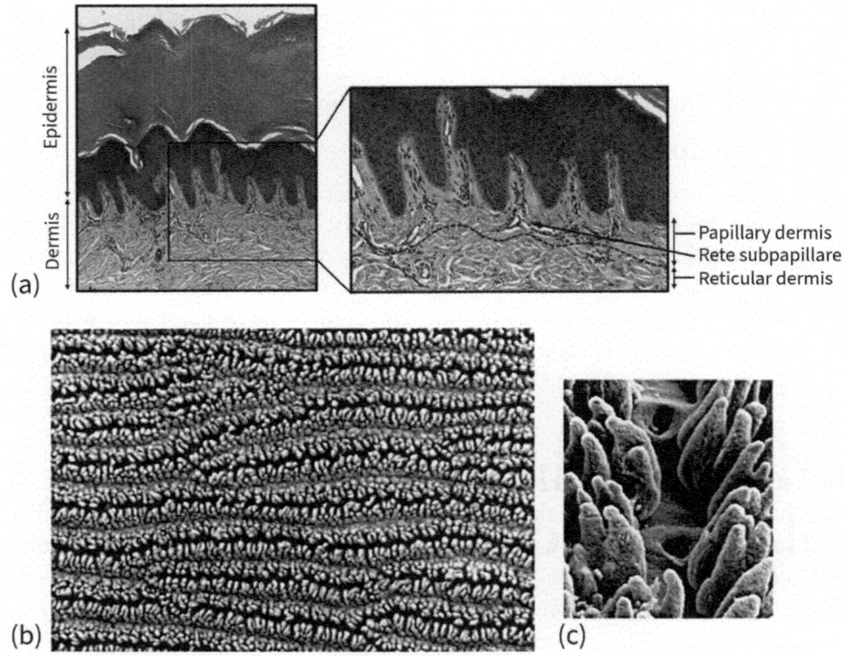

FIGURE 2.2 The microstructural features of the fingertip skin. (a) The adult human skin comprising the epidermis and dermis. The dermis is categorized into papillary and reticular layers, which are delineated by a vascular network known as the rete subpapillare. Reproduced from Sriram et al. [20], with permission. ©2015 Elsevier. All rights reserved. (b) Scanning electron microscopy (SEM) images of the dermal papillae after the removal of the epidermal layer. (c) The conical-shaped dermal papillae magnified 22 times. The craters are openings of tubular channels formerly occupied by ducts of sweat glands. Reproduced from Kawabe et al. [21], with permission. ©1985 Wiley. All rights reserved.

2.2.2.5 Papillary ridges

The surface of the fingertips at the volar skin is characterized by finely organized structures of ridges and grooves (Fig. 2.2a). These papillary ridges are also known as the fingerprint lines. These wavy features project further down into the papillary layer of the dermis. The epidermis and dermis layers interlock at their junction with a topography that can be depicted as a series of peaks and valleys. Fig. 2.2b and 2.2c show the top and magnified views of the dermal papillae after the removal of the epidermal layer.

The microstructural features of the human fingertip skin may be able to offer functional insights useful for replication in artificial tactile sensing devices. Cauna [22] observed that the papillary ridge in the skin surface and the intermediate ridge at the subsurface have a functional relationship, prompting him to describe this microstructure as analogous to mechanical levers. He suggested that the external indentation applied to the fingerprints is magnified by the intermediate ridge acting as a lever.

The mechanical lever theory seems plausible when investigating the human capacity to detect the presence of a single, micro-sized asperity on a very smooth background. As the fingertip is slid across the surface, the local lateral deformation of the papillary ridges was suggested to be responsible for the detection of the small asperities [23]. They suggested that the tactile detection can be accounted for by the activity of the Meissner mechanoreceptors (i.e., FA I end organs) due to their location near the surface and within the intermediate ridges. Furthermore, it was observed that the presence of any asperity could not be perceived during an indentation perpendicular to the surface of the fingertip; i.e., horizontal movements of the skin across the surface were necessary for the fine structures to be perceived.

2.3 Artificial skins

2.3.1 Material composition and characteristics

2.3.1.1 Types of elastomers

Elastomers, derived from the words "elastic" and "polymer" [24], are materials that are primarily used as artificial skins for prosthetic or robotic arms, hands, or fingers. Elastomers can be natural or synthetic rubbers.

Natural rubber is harvested from Hevea trees by making incisions on the trees' barks and collecting the flowing latex sap [25]. These trees can be found in countries with tropical climates like Brazil, Guatemala, India, and Southeast Asian countries like Thailand, Indonesia, Malaysia, and Vietnam. Natural rubbers are composed of polyisoprene, which is a polymer of isoprene (C5H8).

Synthetic rubbers are produced from the polymerization of petroleum-based monomers [28]. Some examples of synthetic rubbers include Styrene-Butadiene Rubber, used as car tires and shoe soles; Nitrile Rubber, used as oil hoses and seals; and Polybutadiene Rubber, used as golf balls. The synthetic rubbers typically used for artificial skins are the silicone and polyurethane types. The scanning electron microscopy (SEM) images of natural rubber and a synthetic rubber of type nitrile rubber are shown in Fig. 2.3.

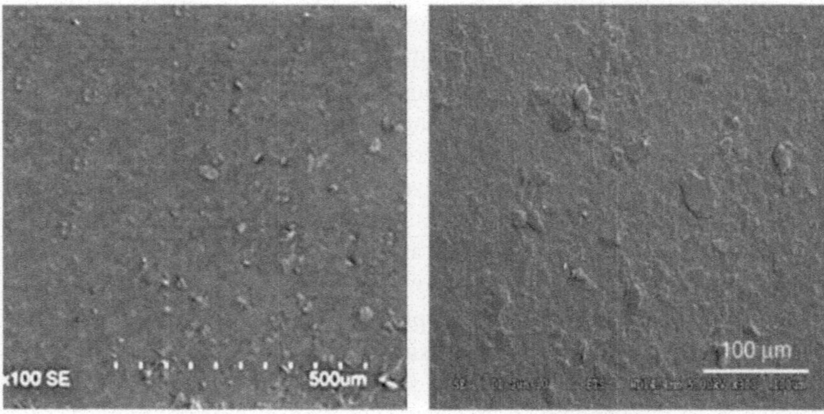

(a) (b)

FIGURE 2.3 Scanning electron microscopy (SEM) imaging. (a) Natural rubber. Reproduced from Jeevakumari et al. [26], with permission. (b) Synthetic rubber, nitrile rubber type. Reproduced from Vinches et al. [27].

2.3.1.2 *Molecular structure*

Natural rubber (Fig. 2.4a) is composed of polyisoprene. It features a sequence of hydrogen (H) and carbon (C) atoms with a double bond between two carbon atoms. Silicone rubber (Fig. 2.4b) is composed of a repeating unit structure, where each silicon (Si) atom is bonded to an oxygen atom and two methyl (CH3) groups to form a flexible and stable polymer chain.

Rubbers are characterized by their amorphous structure, i.e., structures without a regular, crystalline lattice [29]. This characteristic enables rubbers to be elastic, allowing significant stretchability and recovery to the material's original form after the removal of the contacting load. Moreover, the intermittent

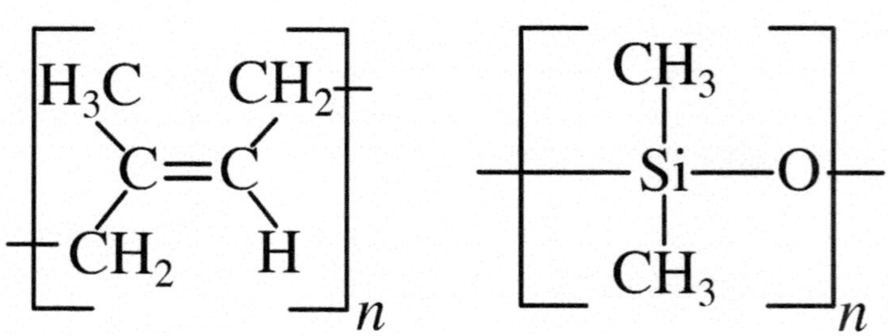

(a) (b)

FIGURE 2.4 Molecular structures of polyisoprene and polydimethylsiloxane (PDMS). (a) Repeating unit of polyisoprene, which is the main component of natural rubber. (b) Repeating unit of polydimethylsiloxane (PDMS), which is a silicone rubber material.

TABLE 2.1 Artificial Skin Materials for Robotics and Prosthetics.

Authors	Year	Application	Material Type	Material	Ref.
Kulkarni et al.	2025	Soft robotics	Silicone	Smooth-Sil and Clear flex, Smooth-On	[30]
Teyssier et al.	2021	Robot skin covering	Silicone	Dragon Skin and Ecoflex, Smooth-On	[31]
Alkhatib et al.	2019	Finger joints	Polyurethane	Poly 74-45, Polytek	[32]
Li et al.	2019	Self-healing skin	Silicone	PDMS	[33]
Rocha et al.	2018	Pressure and bend sensor	Silicone	PDMS	[34]
Cabibihan et al.	2015	Prosthetic hand	Silicone	Ecoflex, Smooth-On	[35]
Cabibihan et al.	2009	Fingertip skin	Silicone	GLS 40, Prochima	[36]
Cabibihan et al.	2006	Fingertip skin	Polyurethane	Poly 74-45, Polytek	[37]

cross-links between the polymer chains enable durability and stability of the material.

A synthetic rubber's ability to replicate the viscoelastic property (i.e., combined viscous and elastic behavior) of the human skin makes it an ideal artificial skin material. This characteristic permits energy absorption and dissipation during contact interactions between an artificial hand and an object.

2.3.1.3 Typical artificial skin materials

Table 2.1 presents a summary of several materials employed in creating artificial skins for humanoid robotics and prosthetics, with a specific emphasis on synthetic rubbers such as silicone and polyurethane. In these works, the materials were selected because of their elasticity, tensile strength, abrasion resistance, and biocompatibility. These attributes were useful in replicating some aspects of the mechanical and sensory properties of the human skin. Further details on the production process, additional compounds used, and testing of the desired mechanical properties can be found in the references mentioned in the table.

2.3.2 Mechanical properties

2.3.2.1 Elasticity

Rubber is composed of long flexible molecules that exhibit constant movement at typical temperatures as a result of thermal agitation [28]. The molecules assume a multitude of random configurations. When force is applied, those molecules undergo straightening and thereafter return immediately to their original random shape as permitted by their thermal motion. This property is the source of rubber's distinctive ability to undergo elastic deformations and fully recover. Rubber molecules have the ability to stretch, but when there is no external force acting on them, they take on compact and random configurations.

Crosslinking or vulcanization is responsible for imparting rubber with its permanent shape since it involves the formation of a limited number of chemical bonds that bind the rubber molecules together [28]. While the molecular strands at the interlinking sites continue to change shape, their movements are now limited because the crosslinks are fixed in place. Before crosslinking, rubber is a viscous and elastic liquid. After crosslinking, rubber transforms into a soft and flexible solid state.

The linear elastic property of rubber can be described by its stress vs. strain relationship:

$$\sigma_x = E\epsilon_x \tag{2.1}$$

where E is known as the Young's modulus, σ_x is the force per unit area along the x-axis, and ϵ_x is the relative strain, which is the deformation along the x-axis in response to the applied force. When a part is subjected to compressive or tensile loads, the strains in the y- and z-axis directions are given as:

$$\epsilon_y = \epsilon_z = -\nu\epsilon_x \tag{2.2}$$

where ν is known as the Poisson's ratio. Rubber is known to have isotropic properties whereby its characteristics are the same in all directions.

When a rubber part is compressed, its volume does not decrease significantly unless the pressure is very high. Consequently, the rubber part bulges sideways and the volume remains the same. Rubbers are considered incompressible with the Poisson's ratio being equivalent to 0.5. With the Young's modulus and Poisson's ratio obtained, the shear modulus G can be calculated as:

$$G = \frac{E}{2(1+\nu)}. \tag{2.3}$$

The linear elastic approximation is applicable to rubber materials under conditions of low strain.

2.3.2.2 *Viscoelasticity*

Rubbers have properties between an ideal linear elastic solid and an ideal viscous liquid. An ideal linear elastic solid obeys Hooke's law where stress is proportional to strain. An ideal viscous liquid obeys Newton's law where stress is proportional to the rate of change of strain with time. Materials that exhibit characteristics between these two extremes are known to have viscoelastic behaviors.

Hooke's law is given as:

$$F = kx \tag{2.4}$$

where F is force and k is the spring constant. Solid materials that obey Hooke's law are known as Hookean solids. They are modeled as springs. Newton's law of viscosity is given as:

$$F = \eta \frac{dx}{dt} \tag{2.5}$$

where η is a viscous damping coefficient and dx/dt is the change of displacement over the change in time (i.e., velocity). Materials that exhibit Newtonian viscous behavior are modeled as dashpots.

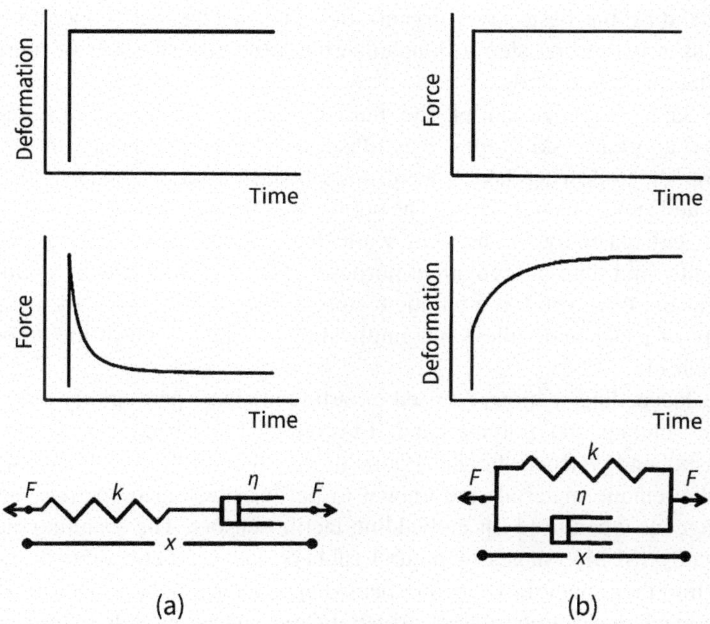

FIGURE 2.5 Mechanical behaviors of artificial skin materials. (a) Maxwell and (b) Kelvin-Voigt models associated with the relaxation and creep behavior of materials, respectively. Adapted from Korhonen and Saarakkala [38]. ©2011 Korhonen and Saarakkala under the terms of the Creative Commons license (CC BY-SA 3.0).

2.3.2.3 Relaxation and creep

A spring arranged in series with a dashpot (i.e., a Maxwell model) and a spring in parallel with a dashpot (i.e., a Kelvin-Voigt model) are models used to describe solid-like elastic behaviors and fluid-like viscous behaviors [39]. These dynamics models are known to describe linear viscoelasticity and are used to show the relaxation and creep behavior of materials.

When a material is subjected to constant deformation, the spring in the Maxwell model results in a stress response, followed by the dashpot's response of an exponential stress relaxation (Fig. 2.5a). No immediate deformation can be observed in a Kelvin-Voigt model for a constant load. However, over time, deformation occurs in an exponential manner (Fig. 2.5b).

The Maxwell model is a good representation of a material's relaxation behavior whereby a material relaxes over time even when the deformation (strain) is kept constant. The Kelvin-Voigt model sufficiently represents a material's creep behavior whereby the material increases its deformation over time even when there is no further increase or decrease of the applied force.

2.4 Summary

Understanding the basic layer organization and mechanical properties of the human skin is the first step to humanize the artificial soft skins of humanoid robots and prosthetic arms.

The layer organization of the human skin, comprising both glabrous (hairless) and hairy skin types, was illustrated. From the surface to the bone structure, the following layers were described: stratum corneum, epidermis, dermis, and subcutaneous tissue. The dermis was found to provide the structural support and elasticity in the skin while the subcutaneous layer provides a cushioning and thermoregulation purpose. The papillary ridge, commonly known as the fingerprint, is a unique feature of the glabrous skin. Its distinctive structure of peaks and valleys has implications in the design of the embedded tactile sensors.

The succeeding sections focused on artificial skins, particularly elastomers such as silicone and polyurethane. Elastomers have been found to mimic the elastic and viscoelastic behaviors of the human skin. Certain silicone and polyurethane materials are known to be durable, biocompatible, and are advantageous when used for embedding tactile sensors. The scanning electron microscopy (SEM) images of natural rubbers and synthetic rubbers gave us insight into these materials' uncomplicated appearance at the micrometer level.

The mechanical properties of artificial skins can be modeled and analyzed using Maxwell and Kelvin-Voigt models to understand behaviors such as stress relaxation and creep. These models can help explain an artificial skin's capability to absorb and dissipate energy upon contact with the external environment.

References

[1] Y. Lanir, Skin mechanics, in: R. Skalak, S. Chien (Eds.), Handbook of Bioengineering, McGraw-Hill, New York, 1987, pp. 1–25.

[2] G. Limbert, Mathematical and computational modelling of skin biophysics: a review, Proceedings of the Royal Society A. Mathematical, Physical and Engineering Sciences 473 (2017) 20170257, https://doi.org/10.1098/rspa.2017.0257.

[3] S.J. Mostafavi Yazdi, J. Baqersad, Mechanical modeling and characterization of human skin: a review, Journal of Biomechanics 130 (2022) 110864, https://doi.org/10.1016/j.jbiomech.2021.110864.

[4] W. Li, Modelling methods for in vitro biomechanical properties of the skin: a review, Biomedical Engineering Letters 5 (2015 Dec) 241–250, https://doi.org/10.1007/s13534-015-0201-3.

[5] A. Buganza Tepole, Computational systems mechanobiology of wound healing, Computer Methods in Applied Mechanics and Engineering 314 (2017) 46–70, https://doi.org/10.1016/j.cma.2016.04.034.

[6] E. Proksch, J.M. Brandner, J.M. Jensen, The skin: an indispensable barrier, Experimental Dermatology 17 (2008) 1063–1072, https://doi.org/10.1111/j.1600-0625.2008.00786.x.

[7] K.O. Johnson, The roles and functions of cutaneous mechanoreceptors, Current Opinion in Neurobiology 11 (2001) 455–461, https://doi.org/10.1016/s0959-4388(00)00234-8.

[8] A. Schirmer, I. Croy, R. Ackerley, What are C-tactile afferents and how do they relate to "affective touch"?, Neuroscience and Biobehavioral Reviews 151 (2023) 105236, https://doi.org/10.1016/j.neubiorev.2023.105236.

[9] J.J. Cabibihan, S.S. Chauhan, Physiological responses to affective tele-touch during induced emotional stimuli, IEEE Transactions on Affective Computing 8 (2017) 108–118, https://doi.org/10.1109/taffc.2015.2509985.

[10] G. Wilkes, A. Nguyen, R. Wildnauer, Structure-property relations of human and neonatal rat stratum corneum I. Thermal stability of the crystalline lipid structure as studied by X-ray diffraction and differential thermal analysis, Biochimica et Biophysica Acta 304 (1973) 267–275, https://doi.org/10.1016/0304-4165(73)90244-4.

[11] M. Harkness, R. Harkness, Functional alterations in the mechanical properties of collageneous frameworks, in: S. Jackson, R. Harkness, S. Partridge, G. Trisran (Eds.), Structure and Function of Connective and Skeletal Tissue, McGraw-Hill, London, 1965, pp. 376–381.

[12] D.T.V. Pawluk, R.D. Howe, Dynamic lumped element response of the human fingerpad, Journal of Biomechanical Engineering 121 (1999 Apr) 178–183, https://doi.org/10.1115/1.2835100.

[13] P.F.D. Naylor, The skin surface and friction, British Journal of Dermatology 67 (1955) 239–248, https://doi.org/10.1111/j.1365-2133.1955.tb12729.x.

[14] R. Tregear, Physical Functions of the Skin, Academic Press, 1966.

[15] A. Viidik, A rheological model for uncalcified parallel-fibred collageneous tissue, Journal of Biomechanics 1 (1968) 3–11, https://doi.org/10.1016/0021-9290(68)90032-8.

[16] R.C. Haut, R.W. Little, A constitutive equation for collagen fibers, Journal of Biomechanics 5 (1972) 423–430, https://doi.org/10.1016/0021-9290(72)90001-2.

[17] R.W. Carton, J. Dainauskas, J.W. Clark, Elastic properties of single elastic fibers, Journal of Applied Physiology 17 (1962) 547–551, https://doi.org/10.1152/jappl.1962.17.3.547.

[18] E.R. Serina, E. Mockensturm, C. Mote, D. Rempel, A structural model of the forced compression of the fingertip pulp, Journal of Biomechanics 31 (1998) 639–646, https://doi.org/10.1016/s0021-9290(98)00067-0.

[19] J. Wu, R. Dong, S. Rakheja, A. Schopper, W. Smutz, A structural fingertip model for simulating of the biomechanics of tactile sensation, Medical Engineering & Physics 26 (2004) 165–175, https://doi.org/10.1016/j.medengphy.2003.09.004.

[20] G. Sriram, P.L. Bigliardi, M. Bigliardi-Qi, Fibroblast heterogeneity and its implications for engineering organotypic skin models in vitro, European Journal of Cell Biology 94 (2015) 483–512, https://doi.org/10.1016/j.ejcb.2015.08.001.

[21] T.T. Kawabe, D.K. Maccallum, J.H. Lillie, Variation in basement membrane topography in human thick skin, The Anatomical Record 211 (1985 Feb) 142–148, https://doi.org/10.1002/ar.1092110205.

[22] N. Cauna, Nature and functions of the papillary ridges of the digital skin, The Anatomical Record 119 (1954) 449–468, https://doi.org/10.1002/ar.1091190405.

[23] R.S. Johansson, R.H. Lamotte, Tactile detection thresholds for a single asperity on an otherwise smooth surface, Somatosensory Research 1 (1983) 21–31, https://doi.org/10.3109/07367228309144538.

[24] F.W. Billmeyer, Textbook of Polymer Science, 3rd edn., John Wiley & Sons, Nashville, TN, 1984 Mar.

[25] S. Thomas, C. Han Chan, L. Pothen, K. Rajisha, H. Maria (Eds.), Natural Rubber Materials, RSC Polymer Chemistry Series, Royal Society of Chemistry, Cambridge, England, 2014 Dec.

[26] S.A. Amutha Jeevakumari, K. Indhumathi, V.R. Arun Prakash, Role of cobalt nanowire and graphene nanoplatelet on microwave shielding behavior of natural rubber composite in high frequency bands, Polymer Composites 41 (2020) 4362–4371, https://doi.org/10.1002/pc.25718.

[27] L. Vinches, N. Testori, P. Dolez, G. Perron, K.J. Wilkinson, S. Hallé, Experimental evaluation of the penetration of TiO2 nanoparticles through protective clothing and gloves under conditions simulating occupational use, Nanoscience Methods 2 (2013) 1–15, https://doi.org/10.1080/21642311.2013.771840.

[28] A.N. Gent, Engineering with Rubber: How to Design Rubber Components, Carl Hanser Verlag GmbH, 2012 Feb.

[29] G.A. Popa (Ed.), Rubber: Types, Properties and Uses, Nova Science, Hauppauge, NY, 2011 Feb.

[30] S.S. Kulkarni, N.M. Bayre, K.A. Khan, Modelling visco-hyperelastic response of silicone based elastomers for soft robotics and foldable structure applications, International Journal of Engineering Science 211 (2025) 104253, https://doi.org/10.1016/j.ijengsci.2025.104253.

[31] M. Teyssier, B. Parilusyan, A. Roudaut, Steimle J. Human-Like, Artificial skin sensor for physical human–robot interaction, in: 2021 IEEE International Conference on Robotics and Automation (ICRA), 2021, pp. 3626–3633, https://doi.org/10.1109/icra48506.2021.9561152.

[32] F. Alkhatib, E. Mahdi, J.J. Cabibihan, Design and analysis of flexible joints for a robust 3D printed prosthetic hand, in: 2019 IEEE 16th International Conference on Rehabilitation Robotics (ICORR), 2019, pp. 784–789, https://doi.org/10.1109/icorr.2019.8779372.

[33] X. Li, R. Yu, H. Yangyang, X. Yang, X. Zhao, Wei H. Self-Healing, Polyurethane elastomers based on a disulfide bond by digital light processing 3D printing, ACS Macro Letters 8 (2019) 1511–1516, https://doi.org/10.1021/acsmacrolett.9b00766.

[34] R.P. Rocha, P.A. Lopes, A.T. de Almeida, M. Tavakoli, C. Majidi, Fabrication and characterization of bending and pressure sensors for a soft prosthetic hand, Journal of Micromechanics and Microengineering 28 (2018) 034001, https://doi.org/10.1088/1361-6439/aaa1d8.

[35] J.J. Cabibihan, D. Joshi, Y.M. Srinivasa, M.A. Chan, Muruganantham A. Illusory, Sense of human touch from a warm and soft artificial hand, IEEE Transactions on Neural Systems and Rehabilitation Engineering 23 (2015) 517–527, https://doi.org/10.1109/tnsre.2014.2360533.

[36] J.J. Cabibihan, S. Pattofatto, M. Jomaa, A. Benallal, M.C. Carrozza, Towards humanlike social touch for sociable robotics and prosthetics: comparisons on the compliance, conformance and hysteresis of synthetic and human fingertip skins, International Journal of Social Robotics 1 (2009) 29–40, https://doi.org/10.1007/s12369-008-0008-9.

[37] J.J. Cabibihan, M.C. Carrozza, P. Dario, S. Pattofatto, M. Jomaa, A. Benallal, The uncanny valley and the search for human skin-like materials for a prosthetic fingertip, in: 2006 6th IEEE-RAS International Conference on Humanoid Robots, 2006, pp. 474–477, https://doi.org/10.1109/ichr.2006.321315.

[38] R.K. Korhonen, S. Saarakkala, Biomechanics and modeling of skeletal soft tissues, in: V. Klika (Ed.), Theoretical Biomechanics, IntechOpen, London, 2011, Chap. 6, https://doi.org/10.5772/19975.

[39] G. Holzapfel, Nonlinear Solid Mechanics, A Continuum Approach for Engineering, John Wiley & Sons, 2000, second print.

Chapter 3

Artificial skin materials modeling

3.1 Introduction

The human skin exhibits complex mechanical behaviors. Such behaviors allow it to distribute contact, conform to surfaces, and provide tactile feedback, while protecting the embedded sensors and structures. To achieve the functional and social requirements of robotic or prosthetic hands, the artificial skins must replicate some of the mechanical properties of the human skin. Understanding the mathematical models of human and soft artificial skins becomes important.

For robotic skins, early models relied on linear elastic assumptions. In the pioneering work by Fearing and Hollerbach [1] for robotic touch, they calculated the resulting stresses and strains experienced by a half-space elastic robotic finger given various contact cases such as point load, line load (similar to a knife's edge), a vertex (a corner), and a planar surface. The models were not sufficient to estimate contact from dynamic loads, considering that the human skin tissue has nonlinear, time-dependent response during contact.

In the study of the human skin, the use of Finite Element Modeling (FEM) has significantly improved the simulations of the fingertip skin. In Wu et al. [2][3], they developed a 2D fingertip FE model, which included the bone, nail, skin layers (epidermis and dermis), and subcutaneous tissue. Hyperelastic and viscoelastic constitutive equations were implemented for the skin layers and subcutaneous tissue. In Dandekar et al. [4], they developed a 3D FE model of human and monkey fingertips. They showed that a multilayered model would be able to represent in vivo data on skin surface displacements.

Improved modeling techniques have been proposed to capture the nonlinear, viscoelastic, and anisotropic behavior (i.e., directionally-dependent property) of the human skin. In Wang and Hayward [5], they proposed an in vivo study of the human fingerpad skin that was subjected to tangential traction. They identified variations in the direction of stiffness as well as hysteretic losses.

Researchers have made efforts to replicate the multilayered design for achieving the mechanical behavior and tactile sensitivities of the human fingertip skin. For example, Osborn et al. [6] and Teyssier et al. [7] designed artificial skins that mimicked the distribution of tactile sensors beneath artificial skins that have viscoelastic properties. In Berselli et al. [8], they investigated the design of fluid-filled robotic fingers to tune the viscoelastic properties of

Humanizing Artificial Touch. https://doi.org/10.1016/B978-0-32-395519-5.00013-1

the artificial skin. This design was able to replicate the human finger's dynamic properties during its interaction with objects.

In the early days, the design of a tactile sensing system proceeded with little consideration of the artificial skin covering to embed the sensors. In Cabibihan et al. [9][10], we proposed to investigate the skin material first prior to embedding the tactile sensors. Some of the needed characteristics include the skin's compatibility with the resulting signals from the embedded sensors, softness, manufacturability, among others. The possibility of simulating the internal effects from external contacts afforded by the finite element methodology enables us to better select the artificial skin material that can satisfy multiple design requirements.

The next section, Sec. 3.2, describes the earlier studies on the constitutive equations for human skin modeling. Sec. 3.3 describes the constitutive equations to model the hyperelastic and viscoelastic behaviors of soft materials for artificial skins. Sec. 3.4 presents the methodology to identify the parameters to describe the test samples. Sec. 3.5 describes the methodology that was used for FE analysis. Sec. 3.6 shows the methodology to validate the simulation results. The last section, Sec. 3.7, summarizes the chapter.

3.2 Human skin modeling

Drawing from biomechanical studies, the skin of the human fingertip can be modeled according to the characteristics of each layer of the skin. A relationship between stress, strain, and deformation in a material can be defined through a mathematical description, known as constitutive equations, where each layer's mechanical behavior can be described under various loading conditions. In the succeeding sections, the constitutive equations for the skin near the surface and the subcutaneous tissue, which is near the supporting bone structure, will be described.

3.2.1 Constitutive equations for the skin

Wu et al. [2] modeled the epidermis and dermis layers of the human skin tissue to have both nonlinear elastic and viscoelastic behaviors. As such, the total stress on the human skin tissue ($\boldsymbol{\sigma}_{skn}$) was proposed to be the combined stress of the elastic ($\boldsymbol{\sigma}_E$) and viscous ($\boldsymbol{\sigma}_V$) stress components:

$$\begin{aligned}\boldsymbol{\sigma}_{skn}(t) &= \boldsymbol{\sigma}_E(t) + \boldsymbol{\sigma}_V(t)\\ &= \boldsymbol{\sigma}_E(t) + \int_0^t \frac{\dot{G}(\tau)}{G_0}\boldsymbol{\sigma}_E(t-\tau)d\tau\\ &= \boldsymbol{\sigma}_E(t) + \int_0^t \dot{g}(\tau)\boldsymbol{\sigma}_E(t-\tau)d\tau,\end{aligned} \tag{3.1}$$

where t is the time. The stress relaxation function was defined using the Prony series [11]:

$$g(t) = \frac{G(t)}{G_0} = \left[1 - \sum_{i=1}^{N_G} g_i (1 - e^{-t/\tau_i}) \right], \tag{3.2}$$

where G_0 is the instantaneous modulus and $G(t)$ is the time-dependent modulus. g_i and τ_i are the parameters of the stress relaxation function. The index, i, can have values from 1 to N_G, where N_G is the number of terms used in the stress relaxation function.

Wu et al. [12] further assumed that the epidermis and dermis layers were behaving as a nonlinear elastic material with hyperelastic characteristics, which is according to the finite deformation theory. To describe this behavior, the strain energy density per unit volume model was obtained from Storakers [13], which was an extension from Ogden's equation for incompressible materials [14]:

$$U = \sum_{i=1}^{N} \frac{2\mu_i}{\alpha_i^2} \left[\lambda_1^{\alpha_i} + \lambda_2^{\alpha_i} + \lambda_3^{\alpha_i} - 3 + \frac{1}{\beta} (J^{-\alpha_i \beta} - 1) \right], \tag{3.3}$$

where $J = \lambda_1 \lambda_2 \lambda_3$ is the volume ratio, λ_i (i = 1, ..., N) are the principal stretch ratios, α_i and μ_i are the material parameters. The indices, i, can have values from 1 to N, where N is the number of terms used in U. The term β is related to the Poisson's ratio, ν, through $\beta = \frac{\nu}{1-2\nu}$.

The strain energy function, U, in Eq. 3.3 was used to obtain the elastic stress ($\boldsymbol{\sigma}_E$):

$$\boldsymbol{\sigma}_E = \frac{2}{J} \boldsymbol{F} \frac{\partial U}{\partial \boldsymbol{C}} \boldsymbol{F}^T \tag{3.4}$$

where $\boldsymbol{F}$ is the deformation gradient tensor and $\boldsymbol{C}$ is the right Cauchy-Green deformation tensor.

3.2.2 Constitutive equations for the subcutaneous tissue

Wu et al. [12] modeled the subcutaneous tissue as a porous material that is fully saturated with fluid. They adapted the biphasic model for articular cartilage by Mow et al. [15] and Holmes and Mow [16]. The subcutaneous tissue was assumed to undergo large deformation. The poroelastic behavior was formulated in accordance with the finite deformation theory with deformation-dependent hydraulic permeability. The stress on the subcutaneous tissue, $\boldsymbol{\sigma}_{sub}$, was calculated to have components of stresses in the fluid phase, $\boldsymbol{\sigma}_f$, and solid phase, $\boldsymbol{\sigma}_s$:

$$\begin{aligned} \boldsymbol{\sigma}_{sub} &= \boldsymbol{\sigma}_f + \boldsymbol{\sigma}_s \\ \boldsymbol{\sigma}_f &= -\Phi_f p \mathbf{I}, \\ \boldsymbol{\sigma}_s &= \boldsymbol{\sigma}_E - \Phi_s p \mathbf{I} \end{aligned} \tag{3.5}$$

where p is the fluid pressure, and the instantaneous volume fractions of the fluid and solid phases are denoted as Φ_f and Φ_s, respectively, with the relationship of $\Phi_f + \Phi_s = 1$ being assumed.

Furthermore, the equation of motion governing the tissue is given as:

$$\nabla \cdot \boldsymbol{\sigma}_s + \boldsymbol{\pi}_s = 0, \quad \nabla \cdot \boldsymbol{\sigma}_f + \boldsymbol{\pi}_f = 0 \tag{3.6}$$

where $\boldsymbol{\pi}_f$ and $\boldsymbol{\pi}_s$ are the momentum exchanges, given by

$$\boldsymbol{\pi}_f = -\boldsymbol{\pi}_s = \Phi_s \nabla p + K(\mathbf{v}_s - \mathbf{v}_f) \tag{3.7}$$

where $\mathbf{v}_s$ and $\mathbf{v}_f$ are the velocities of the fluid and solid phases, respectively, K is the diffusive drag constant (i.e., it measures the frictional resistance of the fluid flow through the solid matrix). K is related to the hydraulic permeability, k, by $K = (\Phi_f)^2/k$.

Wu et al. [12] further considered the hydraulic permeability to be deformation-dependent. They adopted the deformation-dependent permeability constitutive equation by Holmes and Mow [16] given as:

$$k = k_0\left(\frac{e}{e_0}\right)^{\kappa} exp\left\{\frac{M}{2}\left[\left(\frac{1+e}{1+e_0}\right)^2 - 1\right]\right\} \tag{3.8}$$

where e_0 is the void ratio in the undeformed state, M and κ are material parameters, k and k_0 are hydraulic permeabilities at the deformed and undeformed state, respectively.

3.3 Artificial skin materials characterization

3.3.1 Hyperelastic and viscoelastic behavior

In Cabibihan et al. [10], we assumed that the artificial skin would have hyperelastic and viscoelastic properties. As a result, the total stress was made equal to the sum of the hyperelastic *(HE)* and viscoelastic *(VE)* stresses:

$$\boldsymbol{\sigma}_{art}(t) = \boldsymbol{\sigma}_{HE}(t) + \boldsymbol{\sigma}_{VE}(t) \tag{3.9}$$

where t is the time.

3.3.2 Hyperelastic model

The artificial skin materials' hyperelastic behavior was described using a strain energy density per unit volume, U, defined in Storakers [13] for highly compressible elastomers. Wu et al. [12][2], used this for a human fingertip model. Cabibihan et al. [9][17] implemented this formulation for an artificial fingertip. The equation is as follows:

$$U=\sum_{i=1}^{N_h}\frac{2\mu_i}{\alpha_i^2}\left[\lambda_1^{\alpha_i}+\lambda_2^{\alpha_i}+\lambda_3^{\alpha_i}-3+\frac{1}{\beta}(J^{-\alpha_i\beta}-1)\right] \tag{3.10}$$

where μ_i (i = 1, ..., N_h) represent the shear moduli, α_i (i = 1, ..., N_h) are dimensionless material properties, λ_i ($i = 1, 2, 3$) are the principal stretch ratios, $J = \lambda_1\lambda_2\lambda_3$ is the volume ratio, and N_h is the number of terms utilized in the strain energy function. The coefficient β defines the degrees of compressibility in the energy function. Its relationship with the Poisson's ratio, ν, is given as:

$$\beta=\frac{\nu}{1-2\nu} \tag{3.11}$$

F and ***C*** are known as the deformation gradient and the right Cauchy-Green deformation tensors, respectively. Assuming that the candidate materials were incompressible, the volume ratio, J, was set equal to unity. The elastic stress was related to the elastic strain energy function by:

$$\boldsymbol{\sigma}_{HE}=\frac{2}{J}\boldsymbol{F}\frac{\partial U}{\partial \boldsymbol{C}}\boldsymbol{F}^T \tag{3.12}$$

where ***F*** is the deformation gradient tensor and ***C*** is the right Cauchy-Green deformation tensors, respectively.

3.3.3 Viscoelastic model

For the viscoelastic stress, a relaxation function $g(t)$ was applied to the hyperelastic stress. In Wu et al. [12][2] and in Cabibihan et al. [9][17], they implemented the viscoelastic model as follows:

$$\boldsymbol{\sigma}_{VE}=\int_0^t \dot{g}(\tau)\boldsymbol{\sigma}_{HE}(t-\tau)d\tau \tag{3.13}$$

To express multiple relaxation time constants, the stress relaxation function $g(t)$ was developed using the Prony series of order N_v, where g_i and τ_i are viscoelastic parameters:

$$g(t)=\left[1-\sum_{i=1}^{N_v} g_i(1-e^{-t/\tau_i})\right] \tag{3.14}$$

For each material type, the number of coefficients to identify is $2(N_h + N_v)$ $+ 1(\nu)$.

3.4 Parameter identification

3.4.1 Material samples

The candidate materials were chosen based on the material softness. The first material selected was silicone (GLS 40, Prochima, s.n.c., Italy). On the material data sheet, the silicone sample has a Shore A value of 11. The second material selected was polyurethane (Poly 74-45, Polytek Devt. Corp., USA). The polyurethane sample has a Shore A value of 45. A lower Shore value denotes a lower resistance to the indentation tester. These materials were used previously in earlier works in robotic fingers and as embedding materials for tactile sensors [18][19].

3.4.2 Methodology

Indentation tests were carried out on a materials testing machine (model MTS 810, MTS Systems Corp, USA). Industrial grease was applied to both the moving cylinder and the fixed plate on a regular basis to reduce friction. When the force reading was 30 mN, the contact between the plates and the specimen was determined. The test was repeated three times for each of the three different specimens.

Both the viscoelastic and hyperelastic equation coefficients were required. Two tests were required to determine these. First, the viscoelastic behavior coefficients g_i and τ_i were determined using relaxation tests at 20, 30, and 40% nominal strain. The specimen was compressed at a rate of 60 mm/s to peak. This resulted in an initial stress that is theoretically equivalent to the strain-dependent hyperelastic stress. To allow the material to relax, the position of the compressing plate was held constant for approximately 60 s, or until the force was assessed to have achieved its asymptotic value. Since the hyperelastic portion of the stress stayed constant during relaxation, the viscoelastic equation coefficients can be derived directly from the relaxation function.

Second, the hyperelastic behavior coefficients μ_i and α_i were determined from compression tests performed at 1, 5, and 10 mm/s loading rates, until 45% of nominal strain was reached (i.e., 5.5 mm final length). These compression rates are too low to reproduce purely hyperelastic behavior, thus allowing the hyperelastic equation coefficients to be identified. Furthermore, the viscoelastic behavior has taken effect due to the slow loading rates. As a result, the identified coefficients from the relaxation function were used in the identification procedure of the hyperelastic equation coefficients.

3.4.3 Results

The combined hyperelastic and viscoelastic behaviors (Eq. 3.9), the hyperelastic behavior alone (Eq. 3.12), and the viscoelastic behavior alone were tested using the viscoelastic and hyperelastic coefficients (Eq. 3.13). For the silicone material, the combined hyperelastic and viscoelastic behavior curves deviated by 1.4% from the pure hyperelastic condition. The polyurethane material's error was 3.9%.

3.4.3.1 Viscoelastic parameters

Figs. 3.1a and 3.1c show the stress relaxation behavior of silicone and polyurethane for strains of 20, 30, and 40%. The experimental and analytical curves exhibit good agreement. An equation with $N_v = 3$, where N_v stands for the number of terms utilized in the relaxation function of Eq. (3.14), may fit the data curves for all strains. The silicone samples were found to be softer than the polyurethane samples. Figs. 3.1a and 3.1c also show that silicone samples achieved the steady state more quickly than the polyurethane samples. Tables 3.1 and 3.2 display the stress relaxation coefficients for g_i and τ_i.

3.4.3.2 Hyperelastic parameters

Both the $N_h = 2$ hyperelastic equation and the $N_v = 3$ stress relaxation equation were successful in providing a good fit for the experimental compression curves

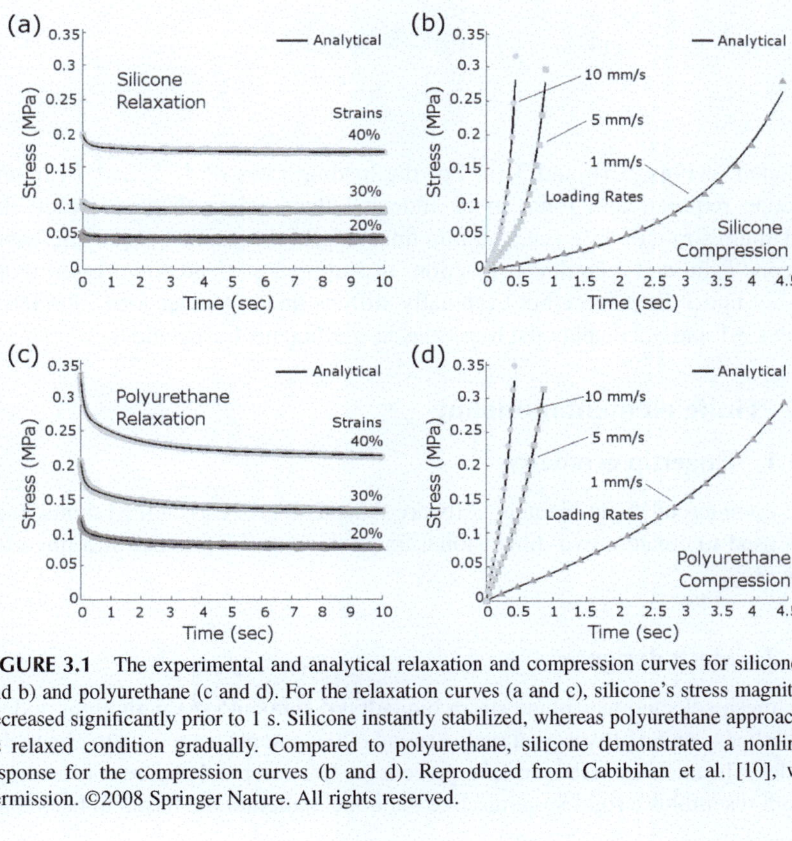

FIGURE 3.1 The experimental and analytical relaxation and compression curves for silicone (a and b) and polyurethane (c and d). For the relaxation curves (a and c), silicone's stress magnitude decreased significantly prior to 1 s. Silicone instantly stabilized, whereas polyurethane approached its relaxed condition gradually. Compared to polyurethane, silicone demonstrated a nonlinear response for the compression curves (b and d). Reproduced from Cabibihan et al. [10], with permission. ©2008 Springer Nature. All rights reserved.

TABLE 3.1 Coefficients for the polyurethane sample ($\nu = 0.47$). Reproduced from Cabibihan et al. [10], with permission.

Parameter	i		
	1	2	3
g_i	0.167	0.158	0.113
τ_i (s)	0.100	1.380	25.472
μ_i (MPa)	0.100	0.063	–
α_i	5.500	8.250	–

TABLE 3.2 Coefficients for silicone sample ($\nu = 0.49$). Reproduced from Cabibihan et al. [10], with permission.

Parameter	i		
	1	2	3
g_i	0.015	0.044	0.029
τ_i (s)	0.025	0.150	0.300
μ_i (MPa)	0.080	0.150	–
α_i	0.001	15.500	–

depicted in Figs. 3.1b and 3.1d. For the loading rates of 1, 5, and 10 mm/s, silicone required less force to be indented than polyurethane to reach the maximum strain of 45% (i.e., 5.5 mm final length). In contrast to polyurethane, silicone exhibited a nonlinear reaction. In addition, silicone compressed more readily under light force but eventually stiffens under heavier load. Similarly, Tables 3.1 and 3.2 display the hyperelastic coefficients for μ_i and α_i.

3.5 Finite element modeling

3.5.1 Fingertip geometry

The commercial finite element software Abaqus FEA (Dassault Systems, Inc) was used to create a two-dimensional finite element (FE) model of a fingertip (Fig. 3.2).

3.5.2 Mesh design

The mesh selected was plane strain biquadratic element type with eight nodes. A thickness unit of 1 was assumed. Around 670 elements were used to create the model. The anchors in the model represent the fixation in all degrees of freedom, which resembles a rigid structure like the bone, nail, and the tissue in between.

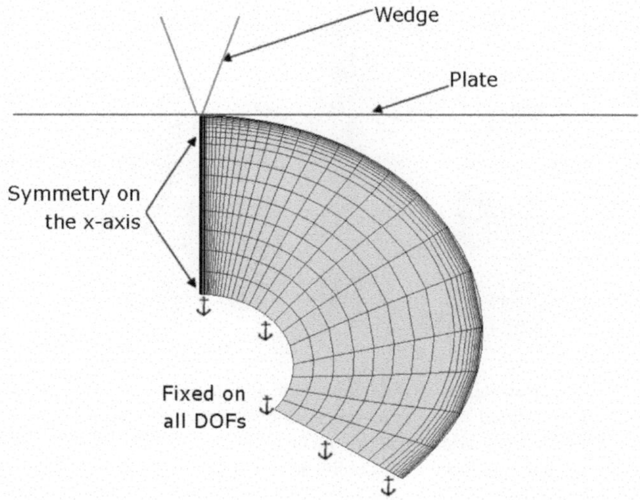

FIGURE 3.2 The synthetic fingertip's two-dimensional finite element model. The shape was designed to resemble a cross-section of a human fingertip. The fixed region functions similarly to the bone and nail in providing stiff support. The flat plate and the wedge indenters were presented to the fingertip model as rigid, analytic surfaces. Reproduced from Cabibihan et al. [10], with permission.

3.5.3 Simulation procedure

The indenters were configured to move vertically downward toward the fingertip. There were two types of indenters used: a flat surface (i.e., plate) and a wedge. The plate was indented 1 mm towards the fingertip at a rate of 0.2 mm/s for the quasi-static plate indentation test. For the wedge indentation test, the loading conditions were comparable to Srinivasan's experiments [20], which were subsequently simulated in Abaqus by Wu et al. [2]. The indenter had the same width as the tip of the wedge used in human experiments (i.e., 50 μm) [20]. During the first second, a 1 mm indentation was created and maintained for the next 100 seconds. Both indenters were modeled as a rigid, analytical surface. It was assumed that the contact between the indenter and the fingertip was frictionless.

3.6 Validation experiments

3.6.1 Methodology

The fingertip prototypes were manufactured using the polyurethane and silicone materials. The geometry depicted in Fig. 3.2 was made symmetrical and physically extended to the longitudinal direction by 22 mm. For each of these materials, three prototypes were constructed. Each fingertip prototype was glued (Polyolefin 406, Loctite) to a thermoplastic rigid base material using a 3D

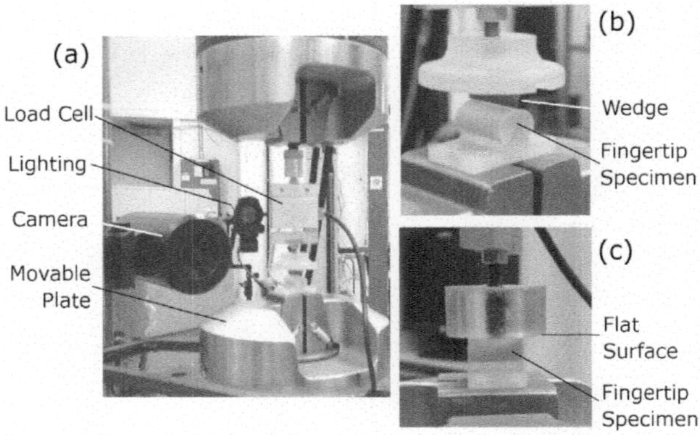

FIGURE 3.3 The experimental setup for compression and imaging to validate the finite element model. (a) The setup consisted of the materials testing machine, a lighting system, and a CCD camera. (b) The wedge indenter is shown above the fingertip specimen, while (c) the flat indenter is shown in a similar setup. Reproduced from Cabibihan et al. [10], with permission. ©2008 Springer Nature. All rights reserved.

printer. The base was assumed to be a single material composed of the bone, nail, and material between them. The Young's modulus for the thermoplastic was 775 MPa.

The experiments were carried out on a materials testing machine. A lighting system and a CCD camera capable of capturing images with a resolution of 1280 × 1024 pixels were also included. The indenters were attached to a load cell, which was then mounted on the testing machine's fixed part (Fig. 3.3).

The fingertip sample was indented 1 mm towards the fixed wedge at a rate of 1 mm/s for the wedge indentation test. The displacement was kept constant for 100 s. Images were taken at the start of the experiment ($T = 0$ s). Additional images were taken at 1 s and at 100 s. The test specimen was indented with a displacement of 1 mm at a rate of 0.2 mm/s for the quasi-static loading test. Imaging experiments were conducted using a 50 μm fine wedge tip. Every time an experiment was run, a different sample was used to allow recovery from any relaxation-related effects. Each material underwent five different tests.

In addition to comparing the force-time-displacement curves, it was also possible to compare the simulation's surface profile with the outcomes of the experiment. The acquired images were digitally post-processed. A three-step process was used to extract the surface profile of the finger prototypes. The reference point in the image had to be identified first. To achieve this, the image at time $T = 0$ was superimposed on images at time $T = 1$ or $T = 100$ (Fig. 3.4). The intersection of the wedge's right corner and the $T = 0$ s curve served as the reference point. Second, a resolution of 10 μm was chosen for the grid size on the digitizer. To increase accuracy once the profiles were to be digitized, the grids that coincided with the fingertip profile were marked with crosshairs. Finally, using a code written in Matlab®, the crosshairs representing the fingertip

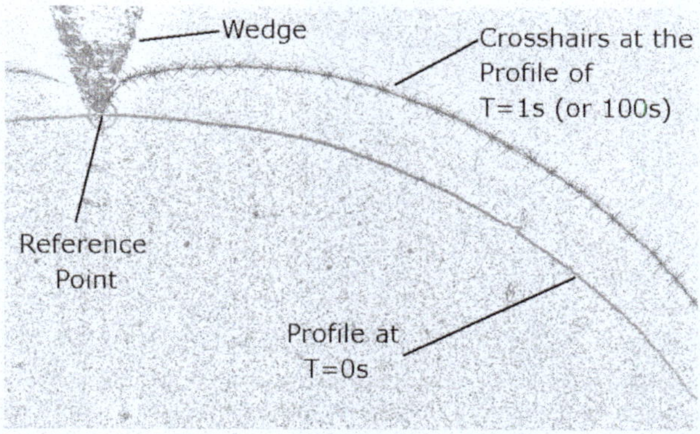

FIGURE 3.4 Digital data extraction of the fingertip surface profile. To establish a reference point, the picture at $T = 0$ s (lower curve) was superimposed with the images at $T = 1$ s and $T = 100$ s (upper curve). The reference point was established at $T = 0$ s at the junction of the wedge's rightmost corner and the fingertip's profile. Crosshairs were placed on the points of the top curve's profile that coincided with the 10 μm resolution grids of an image processing software (e.g. GIMP, Photoshop) in order to make manual digitizing easier and more accurate. Reproduced from Cabibihan et al. [10], with permission.

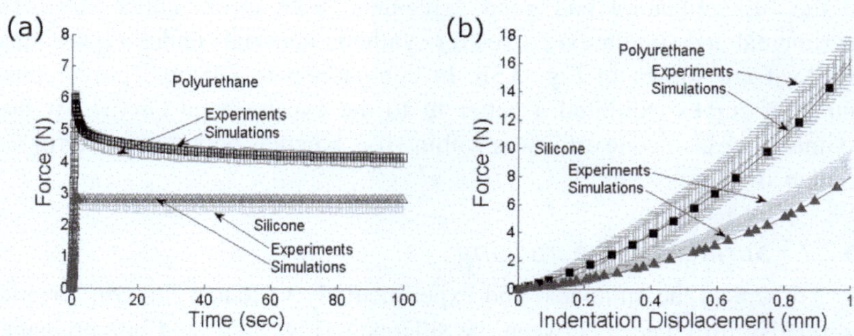

FIGURE 3.5 Validation of the wedge (a) and plate indentations (b). (a) To validate the FE fingertip model, the force-time curves in the wedge indentation experiment and simulations were compared for silicone and polyurethane materials. (b) For the silicone and polyurethane materials, the force-displacement curves in the plate indentation experiment and simulations were compared. Reproduced from Cabibihan et al. [10], with permission.

surface profiles at $T = 1$ s and at $T = 100$ s were manually digitized and plotted separately.

3.6.2 Results

3.6.2.1 Indentation and compression tests

The results of the FE simulation for the wedge indentation were presented in Fig. 3.5a using the mean and standard deviation of the five tests for each material. Silicone was softer than polyurethane because it could be indented

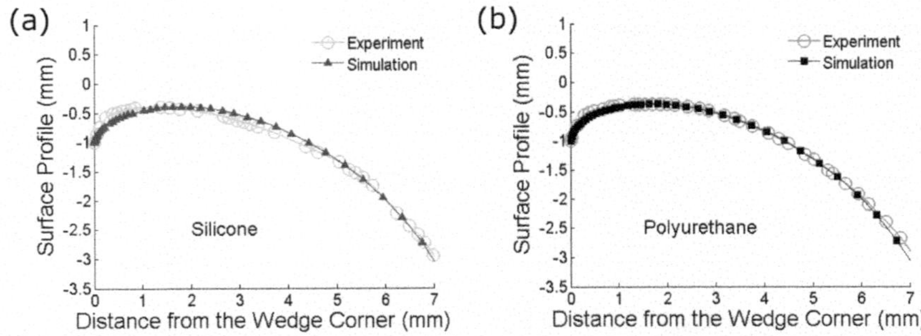

FIGURE 3.6 Surface profiles of synthetic fingertips for (a) silicone and (b) polyurethane simulation and validation. The surface profiles were obtained from the wedge's rightmost corner to a distance of 7 mm (the full width was 8 mm). Reproduced from Cabibihan et al. [10], with permission. ©2008 Springer Nature. All rights reserved.

by a wedge to a depth of 1 mm with less than half the force necessary for polyurethane. In terms of relaxation, silicone recovered rapidly, whereas polyurethane stabilized for another 100 seconds. The experimental curves and the FE simulation had good agreement, with errors under 10%. The experimental and simulation curves for synthetic materials under a quasi-static plate load are shown in Fig. 3.5b. In comparison to silicone, polyurethane simulation curves exhibited a better fit to the experimental curves. At their maximal forces, silicone error was within 10%, whereas polyurethane error was less than 10%.

3.6.2.2 Surface profile imaging

Fig. 3.6 depicts the simulated and experimentally validated fingertip profiles. The relative differences between the silicone's (Fig. 3.6a) and polyurethane's (Fig. 3.6b) simulated curves and the corresponding averaged experimental data are less than 10%. Differences between the simulation and validation results for silicone and polyurethane were expected. As shown in Fig. 3.5, these discrepancies are likely due to the parameter estimation process. The imaging results confirm the simulation results, which made use of the constitutive equations and the parameters obtained from the experiments.

3.7 Summary

Understanding the mechanical behavior of the human fingertip skin, through its biomechanical model, will enable us to replicate the desired properties that we can implement in the artificial fingertip.

This chapter first presented the previous constitutive equations for the external layer of the human skin (i.e., epidermis and dermis layers) and the internal layer (i.e., subcutaneous tissue). The external layer is stiffer for protection while the internal layer is softer for impact absorption.

From those earlier models, it was found that the human skin's hyperelastic and viscoelastic behaviors could be an ideal model for artificial skins. The artificial skins were characterized with combined hyperelastic and viscoelastic behaviors. The hyperelastic model consisted of a strain energy density function and the viscoelastic model consisted of a Prony series relaxation function.

The identification of parameters was carried out using standard compression and relaxation tests. Samples of silicone and polyurethane were tested. Both were soft materials with Shore A 11 and Shore A 45 values, respectively. While the polyurethane samples had higher resistance to compression, the silicone samples relaxed much faster than the polyurethane samples. This finding can be used to select which material can be used for external layer and internal layer.

The next procedure involved the development of a finite element model of the fingertip. To simplify the geometry, a cross-section of a fingertip was used for modeling. The model was further simplified with mirror-symmetry on the central line. The mesh made use of plane strain elements. Fine meshes were implemented on the contacting areas of the fingertip model. The wedge and plate indenters were assumed to have frictionless contact.

To establish validity, the results from the finite element model, with hyper-viscoelastic behaviors and the appropriate parameters defined, were compared with physical test results. The indentation test conditions with the wedge and the plate indenters showed agreement in their force-displacement and force-time curves, with errors under 10%. Additionally, imaging tests showed overlapping surface profile curves for both the simulation and physical tests.

When designing tactile sensing systems, the selection of artificial skins has normally been conducted as a post hoc activity after the tactile sensing system has been developed. This chapter proposed a workflow that can make designers perform simulation studies prior to the costly trial-and-error approach for embedding the tactile sensors.

References

[1] R.S. Fearing, J.M. Hollerbach, Basic solid mechanics for tactile sensing, The International Journal of Robotics Research 4 (1985) 40–54, https://doi.org/10.1177/027836498500400304.

[2] J. Wu, R. Dong, S. Rakheja, A. Schopper, W. Smutz, A structural fingertip model for simulating of the biomechanics of tactile sensation, Medical Engineering & Physics 26 (2004) 165–175, https://doi.org/10.1016/j.medengphy.2003.09.004.

[3] J.Z. Wu, R.G. Dong, Analysis of the contact interactions between fingertips and objects with different surface curvatures, Proceedings of the Institution of Mechanical Engineers. Part H, Journal of Engineering in Medicine 219 (2005) 89–103, https://doi.org/10.1243/095441105x9327.

[4] K. Dandekar, B.I. Raju, M.A. Srinivasan, 3-d finite-element models of human and monkey fingertips to investigate the mechanics of tactile sense, Journal of Biomechanical Engineering 125 (2003) 682–691, https://doi.org/10.1115/1.1613673.

[5] Q. Wang, V. Hayward, In vivo biomechanics of the fingerpad skin under local tangential traction, Journal of Biomechanics 40 (2007) 851–860, https://doi.org/10.1016/j.jbiomech.2006.03.004.

[6] L. Osborn, H. Nguyen, J. Betthauser, R. Kaliki, N. Thakor, Biologically Inspired Multi-Layered Synthetic Skin for Tactile Feedback in Prosthetic Limbs, 2016 38th Annual International Conference of the IEEE Engineering in Medicine and Biology Society (EMBC), IEEE, 2016, pp. 4622–4625.
[7] M. Teyssier, B. Parilusyan, A. Roudaut, Steimle J. Human-Like, Artificial skin sensor for physical human-robot interaction, in: 2021 IEEE International Conference on Robotics and Automation (ICRA), 2021, pp. 3626–3633, https://doi.org/10.1109/icra48506.2021.9561152.
[8] G. Berselli, M. Piccinini, G. Palli, G. Vassura, Engineering design of fluid-filled soft covers for robotic contact interfaces: guidelines, nonlinear modeling, and experimental validation, IEEE Transactions on Robotics 27 (2011) 436–449, https://doi.org/10.1109/tro.2011.2132970.
[9] J.J. Cabibihan, S. Pattofatto, M. Jomaa, A. Benallal, M. Carrozza, P. Dario, The conformance test for robotic/prosthetic fingertip skins, in: The First IEEE/RAS-EMBS International Conference on Biomedical Robotics and Biomechatronics, BioRob 2006, 2006, pp. 561–566, https://doi.org/10.1109/biorob.2006.1639148.
[10] J.J. Cabibihan, S. Pattofatto, M. Jomaa, A. Benallal, M.C. Carrozza, Towards humanlike social touch for sociable robotics and prosthetics: comparisons on the compliance, conformance and hysteresis of synthetic and human fingertip skins, International Journal of Social Robotics 1 (2009) 29–40, https://doi.org/10.1007/s12369-008-0008-9.
[11] N.W. Tschoegl, The Phenomenological Theory of Linear Viscoelastic Behavior, Springer, Berlin, Heidelberg, 1989.
[12] J. Wu, R. Dong, W. Smutz, S. Rakheja, Dynamic interaction between a fingerpad and a flat surface: experiments and analysis, Medical Engineering & Physics 25 (2003 Jun) 397–406, https://doi.org/10.1016/s1350-4533(03)00035-3.
[13] B. Storakers, On material representation and constitutive branching in finite compressible elasticity, Journal of the Mechanics and Physics of Solids 34 (1986) 125–145, https://doi.org/10.1016/0022-5096(86)90033-5.
[14] R.W. Ogden, Large deformation isotropic elasticity: on the correlation of theory and experiment for incompressible rubberlike solids, Proceedings of the Royal Society of London. Series A, Mathematical and Physical Sciences 326 (1972) 565–584, https://doi.org/10.1098/rspa.1972.0026.
[15] V.C. Mow, S.C. Kuei, W.M. Lai, C.G. Armstrong, Biphasic creep and stress relaxation of articular cartilage in compression: theory and experiments, Journal of Biomechanical Engineering 102 (1980 Feb) 73–84, https://doi.org/10.1115/1.3138202.
[16] M. Holmes, V. Mow, The nonlinear characteristics of soft gels and hydrated connective tissues in ultrafiltration, Journal of Biomechanics 23 (1990 Jan) 1145–1156, https://doi.org/10.1016/0021-9290(90)90007-p.
[17] J.J. Cabibihan, M.C. Carrozza, P. Dario, S. Pattofatto, M. Jomaa, A. Benallal, The uncanny valley and the search for human skin-like materials for a prosthetic fingertip, in: 2006 6th IEEE-RAS International Conference on Humanoid Robots, 2006, pp. 474–477, https://doi.org/10.1109/ichr.2006.321315.
[18] B. Edin, L. Ascari, L. Beccai, S. Roccella, J.J. Cabibihan, M. Carrozza, Bio-inspired sensorization of a biomechatronic robot hand for the grasp-and-lift task, Brain Research Bulletin 75 (2008) 785–795, https://doi.org/10.1016/j.brainresbull.2008.01.017.
[19] L. Beccai, S. Roccella, L. Ascari, P. Valdastri, A. Sieber, M.C. Carrozza, P. Dario, Development and experimental analysis of a soft compliant tactile microsensor for anthropomorphic artificial hand, IEEE/ASME Transactions on Mechatronics 13 (2008) 158–168, https://doi.org/10.1109/tmech.2008.918483.
[20] M. Srinivasan, Surface deflection of primate fingertip under line load, Journal of Biomechanics 22 (1989 Jan) 343–349, https://doi.org/10.1016/0021-9290(89)90048-1.

Chapter 4

Artificial skin conformance and compliance

4.1 Introduction

Conformance has been described as the skin's ability to mold and adapt to the surface features of an object, which is critical for activating mechanoreceptors that encode spatial information [1]. Vega-Bermudez and Johnson [1] described conformance as being analogous to the eye's ability to focus an image on the retina. They mentioned that if the skin does not conform to an object's contours, the mechanoreceptors that are embedded underneath the skin cannot accurately transmit the spatial details.

Moreover, Vega-Bermudez and Johnson [1] measured the conformance of the human fingertip skin as the depth of penetration of the skin to the slots of gratings. They measured conformance by assessing how deeply the skin invaded slots of mechanical gratings and directly correlating this metric with tactile acuity. The selected gratings in the experiments had varying widths of 0.35 to 2.0 mm.

They showed that the human fingertip skin with higher conformance (i.e., softer fingertips) exhibited larger contact areas and lower interior stress, enhancing mechanoreceptor activation. They further showed that the anisotropic composition of the fingertip skin has an effect on sensing the direction of the gratings. The gratings that were aligned with the skin ridges (i.e., parallel to the finger axis) were better discriminated than the gratings that were presented in an orthogonal direction to the skin ridges.

Compliance refers to the skin's ability to deform under an applied force, typically quantified by measuring indentation depth under a specific load [1]. Woodward [2] assessed compliance by measuring the indentation produced by a 10 g force.

Skin compliance on human fingertips can be measured in two ways: passively or actively [3][4]. Passive compliance is conducted with the fingerpad being fixated on the test platform while indentation forces are applied to the fingerpad at a constant rate. Active compliance is measured when participants are asked to press their fingertips on compliant objects while forces and displacements are measured.

Humanizing Artificial Touch. https://doi.org/10.1016/B978-0-32-395519-5.00014-3

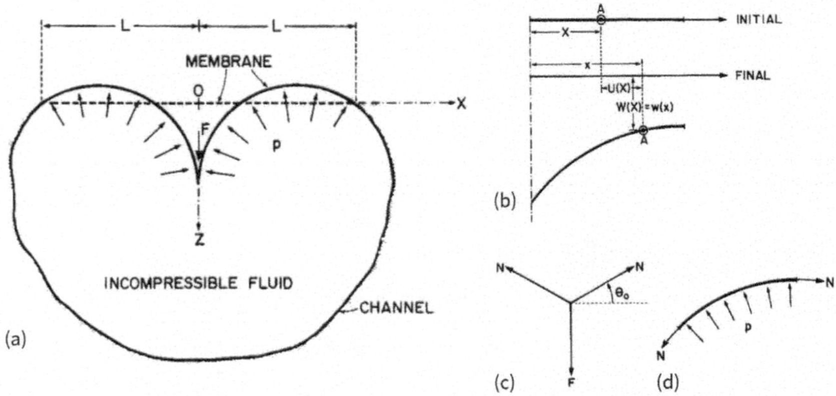

FIGURE 4.1 Waterbed model for analyzing surface deflection. (a) Schematic diagram of the waterbed model of fingertip skin deformation. (b) Before and after diagrams of skin deformation. (c) Free body diagram (FBD) at the location of the load, F. (d) FBD for an element of the membrane in the deformed state. Reproduced from Srinivasan [5], with permission.

The next section is organized according to conformance studies (Sec. 4.2) and compliance studies (Sec. 4.3) in both human and artificial skins. Lastly, Sec. 4.4 summarizes findings on skin conformance and compliance.

4.2 Skin conformance

4.2.1 Human skin conformance

4.2.1.1 Waterbed model

Srinivasan [5] proposed a model to understand the conformance of fingertips during tactile interactions. In the so-called waterbed model, the fingertip was modeled as an elastic membrane filled with an incompressible fluid (Fig. 4.1). The waterbed representation allowed the modeling of large deformations that occur during tactile sensing.

In Fig. 4.1(a), consider an initially flat, linear-elastic membrane with negligible thickness and a width of $2L$. This membrane rests upon an incompressible fluid, which is contained within a rigid-walled channel of arbitrary cross-section. The channel, along with the membrane-covered fluid, is assumed to be in a state of plane strain. The applied load and the resulting membrane stress are expressed as force per unit length, oriented perpendicularly to the cross-section.

Point 0 on the membrane was designated as the origin of a Cartesian coordinate system (X, Z), establishing positive directions. A downward, concentrated load F is applied at point 0. With the assumption of incompressibility of the fluid, the membrane's deformation ensures that the cross-sectional area of the fluid remains constant (i.e., the fluid area above the X-axis equals the area below). This deformation results in a deflected shape of the membrane.

4.2.1.2 Deformations

A particle initially located at $(X, 0)$ on the membrane now occupies a position $(X + U(X), W(X))$, where $U(X)$ and $W(X)$ represent the horizontal and vertical displacements, respectively, expressed as functions of the initial coordinate X (Fig. 4.1b). Let x be defined as the final coordinate, such that $x = X + U(X)$. The vertical displacement is denoted as $w(x)$, which is expressed as a function of the final coordinate x.

It is important to recognize that for any arbitrary X, $W(X) = w(x)$, where $x = X + U(X)$. $W(X)$ and $w(x)$ allow for measurements to be taken from superimposed undeformed and deformed fingertip profiles. With this geometric analysis, the symmetry about the load axis can be used to calculate the deformations.

4.2.1.3 Governing equations

The waterbed model was developed based on the following governing equations. First, the incompressibility constraint was mathematically expressed as:

$$\int_0^L w(x)dx = 0 \tag{4.1}$$

Second, for the equilibrium of the deformed member under the load at $x = 0$ (Fig. 4.1c), the relationship is:

$$F = 2N \sin\theta_0 \tag{4.2}$$

Here, F represents the applied load, and N denotes the membrane force. For $0 < x < L$ (Fig. 4.1d), the equilibrium conditions in the normal and tangential directions relative to the membrane are:

$$N = pR \tag{4.3}$$

where p is the fluid pressure, R indicates the radius of curvature of the membrane, and N is constant.

Third, the constitutive law for the linear-elastic membrane was defined as:

$$N = E\epsilon \tag{4.4}$$

where the membrane stiffness is denoted as E, and the finite strain, ϵ, is defined as:

$$\epsilon = \frac{dU}{dX} + \frac{1}{2}\left[\left(\frac{dU}{dX}\right)^2 + \left(\frac{dW}{dX}\right)^2\right]. \tag{4.5}$$

Lastly, the symmetry condition at $X = 0$ and the fixation of the membrane at $X = L$ set the following conditions:

$$U = 0 \quad \text{at } X = 0 \text{ and } X = L \tag{4.6a}$$

$$W = 0 \quad \text{at } X = L \tag{4.6b}$$

The results from the waterbed model, the experimental data from a human fingertip, and the conformance results from artificial skin materials will be compared in the next section.

4.2.2 Artificial skin conformance

4.2.2.1 Simulation procedure

Simulations were performed to compare the surface deflection of both the skin materials and the human fingertip during the wedge experiments conducted by Srinivasan [5]. Srinivasan conducted experiments wherein a wedge measuring 0.05 mm was applied to the human fingertip and the resulting surface deflection was quantified. In order to conduct a comparison, it was important to evaluate the force required by the wedge to create an indentation of 1 mm on both human fingertips and artificial skin materials.

In order to address this issue, the simulation made use of the findings from Wu et al.'s wedge study [6], which accounted for the force-time characteristics of the fingertip. The findings of the fingertip simulation conducted in reference [6], in conjunction with polyurethane and silicone materials, were subjected to normalization procedures relative to the peak force magnitude exhibited by the polyurethane material.

4.2.2.2 Materials

Two materials were evaluated in this study: silicone (GLS 40, Prochima, Italy) and polyurethane (Poly 74-45, Polytek Devt Corp, USA). The silicone sample has a Shore A value of 11 while the polyurethane sample has a value of 45 (i.e., a lower value indicates a lower resistance to an indenter in a standard durometer test). Three samples were obtained for each of these materials. The samples were prepared in molds having holes of 8 mm diameter and a height of 10 mm.

4.2.2.3 Geometry and mesh design

A two-dimensional finite element (FE) model of a fingertip (Fig. 4.2a) was created in the commercial finite element software Abaqus. The plane strain 8-node bi-quadratic element type was used to model the fingertip. A thickness unit of 1 was set. About 670 elements were used in the model. The interface at the bottom region of the model was fixed in all the degrees of freedom.

4.2.2.4 Loading

The indenters were set to have a vertical downward movement towards the fingertip. For the wedge indentation test, the loading conditions were similar to the experiments of Srinivasan [5], which were later simulated by Wu et al. [6]. The indenter had a width of 0.05 mm similar to the tip of the wedge used in the human experiments [5]. An indentation of 1 mm was applied at the first second and was maintained constant for the next 100 s.

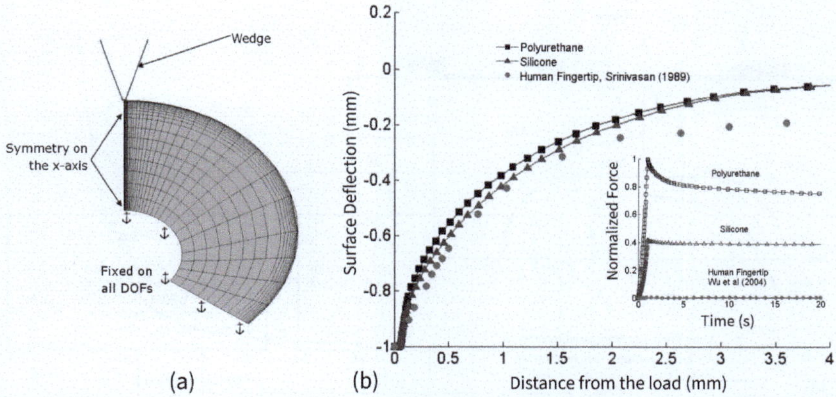

FIGURE 4.2 Simulation model and results of an artificial finger. (a) Geometry and mesh of the fingertip model. A wedge with a width of 50 μm indents the fingertip to 1 mm depth. (b) Indentation results comparing the conformance of the polyurethane and silicone samples with the experimental results from Srinivasan [5]. The inset shows the normalized force comparisons showing the low force needed to indent the human finger to 1 mm depth with the wedge. Reproduced from Cabibihan et al. [7], with permission.

4.2.2.5 Results

The surface deflection of the human fingertip experiments of Srinivasan [5] and the simulation results of the polyurethane and silicone synthetic materials were plotted together on Fig. 4.2b. The synthetic materials had a fairly good overall agreement with the human fingertip from the wedge corner to around 2 mm distance, demonstrating the synthetic materials' capacity to bend or follow surface features. However, the human fingertip requires minimal effort to be indented. The force-time curves from the simulations are shown on Fig. 4.2 (inset) comparing the human fingertip and the synthetic materials. As shown by the resulting plot near zero, the human fingertip conforms to a wedge with very minimal force exerted.

4.3 Skin compliance

4.3.1 Human skin compliance

Serina et al. [8] proposed a structural model to capture the geometric and material complexity of the fingertip to understand the biomechanical behavior during tactile interactions. They modeled the fingertip as a pressurized, axisymmetric ellipsoidal membrane. Such a model represents an incompressible, inviscid fluid which is enclosed by skin. This model can be used to simulate fingertip contact with flat surfaces that is common in a real-world setting.

Fig. 4.3a shows the membrane initially as an ellipsoid with major principal axis length, a_0, set to 10.5 mm and an aspect ratio, $\epsilon = b_0/a_0$, of 0.5, where b_0 is the minor principal axis length. When inflated, the ellipsoid can reach a

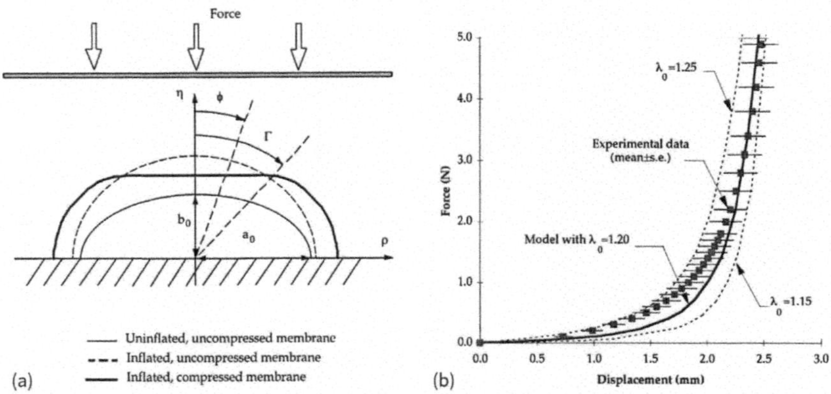

FIGURE 4.3 The forced compression of the fingertip pulp model. (a) The FBD of the fingerpulp in various loaded and unloaded conditions. (b) The theoretical and experimental force-displacement result of the fingertip membrane. Reproduced from Serina et al. [8], with permission.

height, η, of 14 mm and a width, ρ, of 17 mm. These dimensions represent the measurements of human index fingers [9]. The thickness of the membrane, h_0, representing the thickness of the human skin's epidermis was made to be 1 mm. With pressurization, the initial tension in the membrane can simulate the natural tension in the fingertip skin. The stretch at the pole, λ_0, was assigned a value of 1.2.

The model was further assumed to behave according to the membrane theory with the assumption that the membrane is small relative to other dimensions and is axisymmetric. This assumption implied that the profiles of the inflated and compressed membranes are asymmetric and have no shear stresses. The position vectors in the reference, **X**, and deformed, **x**, configurations are:

$$\mathbf{X} = a_0 \sin\phi(\cos\theta\mathbf{e}_1 + \sin\theta\mathbf{e}_2) + b_0\cos\phi\mathbf{e}_3 \tag{4.7}$$

$$\mathbf{x} = \rho(\phi)[\cos\theta\mathbf{e}_1 + \sin\theta\mathbf{e}_2] + \eta(\phi)\mathbf{e}_3, \tag{4.8}$$

where ϕ is measured from the vertical η-axis and $\mathbf{e}_1$, $\mathbf{e}_2$, and $\mathbf{e}_3$ are unit vectors from a fixed Cartesian reference frame and $\mathbf{e}_3$ coincides with the η-axis.

Principal stretch ratios in the meridian, λ_1, and circumferential, λ_2, directions were calculated as the ratio between the lengths of the infinitesimal line elements in the reference, X, and deformed, x, configurations. The subscripts 1 and 2 were used to refer to the meridian and circumferential directions of the membrane, respectively. The prime symbol denotes a derivative with respect to ϕ:

$$\lambda_1(\phi) = \frac{1}{r_0(\phi)}\left(\left|\frac{\rho'(\phi)}{a_0}\right|^2 + \left|\frac{\eta'(\phi)}{a_0}\right|^2\right)^{1/2} \tag{4.9a}$$

$$\lambda_2(\phi) = \frac{\rho(\phi)}{a_0 \sin\phi} \tag{4.9b}$$

where $r_0(\phi) = (a_0^2 \cos^2\phi + b_0^2 \sin^2\phi)^{1/2}$.

Due to membrane incompressibility, the volume must be conserved, and the product of all three principal stretches is unity: $I_3 = \lambda_1^2\lambda_2^2\lambda_3^2 = 1$. This constraint also allows formulation of the strain invariants, I_1 and I_2:

$$I_1 = \lambda_1^2 + \lambda_2^2 + \frac{1}{\lambda_1^2\lambda_2^2}, \tag{4.10a}$$

$$I_2 = \frac{1}{\lambda_1^2} + \frac{1}{\lambda_2^2} + \lambda_1^2\lambda_2^2. \tag{4.10b}$$

Equilibrium in the tangential and normal directions leads to the following equations:

$$\frac{dT_1}{d\rho} + \frac{1}{\rho}(T_1 - T_2) = 0, \tag{4.11a}$$

$$\kappa_1 T_1 + \kappa_2 T_2 = P, \tag{4.11b}$$

where P is the pressure directed to the normal to the ellipsoidal membrane, T_1 and T_2 are the principal stress resultants per unit length of the deformed surface and are related to the principal stretch ratios:

$$T_1 = 2h_0\lambda_3(\lambda_1^2 - \lambda_3^2)\left(\frac{\partial W}{\partial I_1} + \lambda_2^2\frac{\partial W}{\partial I_2}\right), \tag{4.12a}$$

$$T_2 = 2h_0\lambda_3(\lambda_2^2 - \lambda_3^2)\left(\frac{\partial W}{\partial I_1} + \lambda_1^2\frac{\partial W}{\partial I_2}\right), \tag{4.12b}$$

where W is the strain energy function of the membrane material. The principal curvatures are denoted as κ_1 and κ_2:

$$\kappa_1 = \frac{-(\rho'\eta'' - \eta'\rho'')}{(\rho'^2 + \eta'^2)^{3/2}} \tag{4.13a}$$

$$\kappa_2 = \frac{-\eta'}{\rho(\rho'^2 + \eta'^2)^{1/2}}. \tag{4.13b}$$

The constitutive model by Veronda and Westman [10] was used for the membrane:

$$W = 0.0134[e^{4.4(I_1-3)} - 1] - 0.0295(I_2 - 3) \tag{4.14}$$

The inflated volume, $V_{inflate}$, and the compressed volume, $V_{compressed}$, were made equal because the membrane was assumed to be an incompressible fluid:

$$V_{inflated} = 2\pi\int_{\phi=(\pi/2)}^{\phi=0} \rho^2 d\eta = V_{compressed} = 2\pi\int_0^{\eta} \rho^2 d\eta. \tag{4.15}$$

The results from the experimental data in Fig. 4.3b from the human fingertip in Serina et al. [9] and the compliance results from artificial skin materials will be compared in the next section.

4.3.2 Artificial skin compliance

4.3.2.1 Simulation procedure

The simulation methodology implemented a visco-hyperelastic finite element model of an anthropomorphic fingertip using silicone (Shore A 11) and polyurethane (Shore A 45). More details on the simulation procedures were described in Chap. 3. Two internal geometries were studied. The silicone model had an open pocket of 1 mm height (Fig. 4.4a) and the polyurethane model had a 2 mm pocket (Fig. 4.4b). Both open pockets were arc-shaped in appearance [11][12].

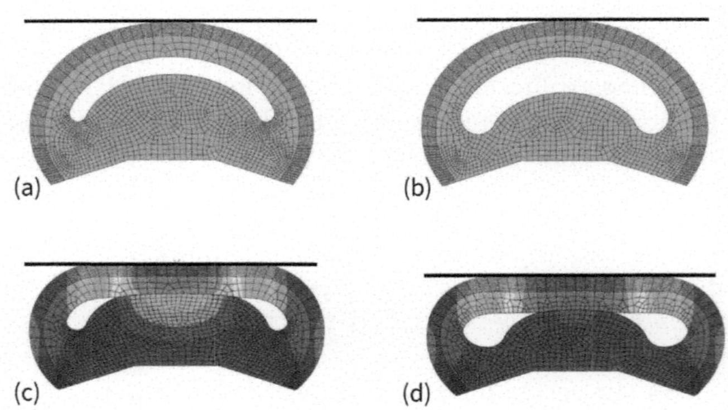

FIGURE 4.4 The finite element model of the fingertip. (a) and (b) The geometries of the 2D finite element fingertip models with silicone and polyurethane behaviors, respectively. (c) and (d) The displacement contour plots of the fingertips when indented with a cyclic loading having 5 N amplitude for silicone and polyurethane, respectively. Reproduced from Cabibihan et al. [13], with permission.

The base region, idealized as the bone and nail, was fully constrained in all degrees of freedom. Boundary and loading conditions reproduced the experimental tapping protocol in Serina [9]. A rigid plate indentation followed the positive half of a sine wave with 5 N amplitude. The tapping frequency was 0.5 Hz. The pulp displacements were recorded at 1 N and 4 N for comparison to human fingertip data. Simulations were executed in a finite element software, and outputs included displacement contours and load–displacement curves with analysis focused on pulp displacements at 1 N and 4 N and the initial stiffness/stiffening behavior versus the human fingertip.

4.3.2.2 Results

The simulations showed that introducing an open pocket substantially increased fingertip compliance (Figs. 4.4c and 4.4d). For the original solid internal geometry, a 1 N plate indentation produced displacements of 0.28 mm (polyurethane) and 0.45 mm (silicone) versus 1.83 mm for the human fingertip. The 1 mm open pocket in the silicone model produced 1.27 mm at 1 N and 1.74 mm at 4 N. The 2 mm open pocket in the polyurethane model produced 1.75 mm at 1 N and 2.33 mm at 4 N. The loading curves for the pocketed designs showed an initial stiffness comparable to the human fingertip and then stiffened as the upper skin layer contacted the solid region fixed to the base, which occurs at approximately 1 mm of displacement for silicone and 2 mm for polyurethane. The comparison results are shown on Fig. 4.5.

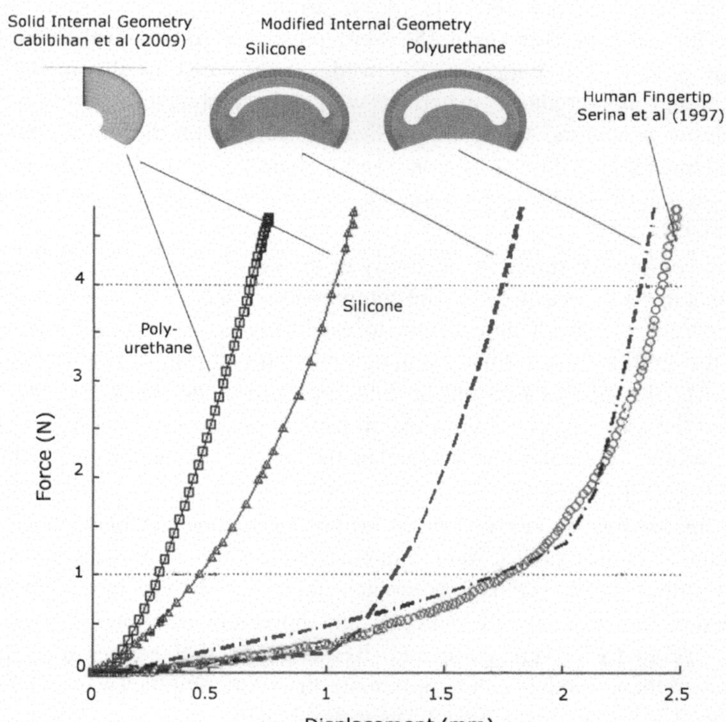

FIGURE 4.5 Comparison of the skin compliance of human and artificial fingertips in various configurations. The solid geometries in Cabibihan et al. [7] with the silicone and polyurethane behaviors can be altered to approach the compliance of a human fingertip in [9] by including internal air pockets on the fingertip geometry. Adapted from Cabibihan et al. [13], with permission.

4.4 Summary

Understanding the stresses and the strains that the fingertip skin experiences when it gets into contact with a surface or an object will allow us to humanize artificial skins. Through the skin conformance and compliance properties that were presented in this chapter, we would be able to replicate the contact interactions with the external environment.

Skin conformance refers to the molding and adaptation of the skin to the surface features of an object [1]. The surface deflections on the fingertip will enable the embedded tactile sensors to be activated to discriminate an object's features. A skin conformance model for a human finger was described by Srinivasan [5]. In this model, the fingertip was analyzed as an elastic membrane that is inflated and filled with incompressible fluid. This model can predict the deflections on the fingertip surface when indented with a wedge.

In the conformance analysis of an artificial finger, the same wedge indentation tests of Srinivasan [5] were prescribed on the validated finite element model of a fingertip. The wedge conformed to the model with silicone and polyurethane materials with the combined hyperelastic and viscoelastic behaviors. While the surface deflections on the human fingertip were comparable with the silicone and polyurethane models, the artificial skin materials required higher forces to achieve the depths prescribed in the experiments.

Skin compliance refers to the ability of the skin to deform due to the force applied at the skin's surface [1]. The compression of the finger due to an applied force recruits the mechanoreceptors to establish stable contact for grasping, shape recognition, and further manipulation of the object. A skin compliance model was described based on the fingertip pulp model of Serina [8]. They modeled the fingertip as a fluid-filled ellipsoidal membrane, which is enclosed by a skin layer. Their model can simulate the contact interactions of the human fingertip with flat surfaces.

In the compliance analysis of an artificial fingertip, the plate indentation experiments of Serina [9] were implemented on the same fingertip model used in the wedge simulations. When the force-displacement curves of the human fingertip and the silicone and polyurethane models, with solid internal geometries, were compared, it was found that neither the silicone nor polyurethane models can approximate the skin compliance of the human fingertip. The polyurethane model was stiffer than the silicone model. However, when the model was hollowed out with an air pocket, even the stiff polyurethane model can approximate the skin compliance of a human fingertip.

A better understanding of the skin conformance and compliance properties enables us to adjust the design of artificial fingertip skins to humanize artificial skins. The modeling and simulation techniques described give us an added tool to compare simulation and experimental results coming from the surface profiles and force-displacement curves of the human and artificial skins.

References

[1] F. Vega-Bermudez, K.O. Johnson, Fingertip skin conformance accounts, in part, for differences in tactile spatial acuity in young subjects, but not for the decline in spatial acuity with aging, Perception & Psychophysics 66 (2004 Jan) 60–67, https://doi.org/10.3758/bf03194861.

[2] K.L. Woodward, The relationship between skin compliance, age, gender, and tactile discriminative thresholds in humans, Somatosensory & Motor Research 10 (1993 Jan) 63–67, https://doi.org/10.3109/08990229309028824.

[3] R.H. LaMotte, Softness discrimination with a tool, Journal of Neurophysiology 83 (2000 Apr) 1777–1786, https://doi.org/10.1152/jn.2000.83.4.1777.

[4] C. Xu, Y. Wang, G.J. Gerling, An elasticity-curvature illusion decouples cutaneous and proprioceptive cues in active exploration of soft objects, PLoS Computational Biology (2021 Mar. 17) e1008848, https://doi.org/10.1371/journal.pcbi.1008848.

[5] M. Srinivasan, Surface deflection of primate fingertip under line load, Journal of Biomechanics 22 (1989 Jan) 343–349, https://doi.org/10.1016/0021-9290(89)90048-1.

[6] J. Wu, R. Dong, S. Rakheja, A. Schopper, W. Smutz, A structural fingertip model for simulating of the biomechanics of tactile sensation, Medical Engineering & Physics 26 (2004) 165–175, https://doi.org/10.1016/j.medengphy.2003.09.004.

[7] J.J. Cabibihan, S. Pattofatto, M. Jomaa, A. Benallal, M.C. Carrozza, Towards humanlike social touch for sociable robotics and prosthetics: comparisons on the compliance, conformance and hysteresis of synthetic and human fingertip skins, International Journal of Social Robotics 1 (2009) 29–40, https://doi.org/10.1007/s12369-008-0008-9.

[8] E.R. Serina, E. Mockensturm, C. Mote, D. Rempel, A structural model of the forced compression of the fingertip pulp, Journal of Biomechanics 31 (1998) 639–646, https://doi.org/10.1016/s0021-9290(98)00067-0.

[9] E.R. Serina, C. Mote, D. Rempel, Force response of the fingertip pulp to repeated compression. Effects of loading rate, loading angle and anthropometry, Journal of Biomechanics 30 (1997 Oct) 1035–1040, https://doi.org/10.1016/s0021-9290(97)00065-1.

[10] D. Veronda, R. Westmann, Mechanical characterization of skin—finite deformations, Journal of Biomechanics 3 (1970 Jan) 111–124, https://doi.org/10.1016/0021-9290(70)90055-2.

[11] J.J. Cabibihan, S.S. Ge, Towards Humanlike Social Touch for Prosthetics and Sociable Robotics: Three-Dimensional Finite Element Simulations of Synthetic Finger Phalanges. Advances in Robotics, Springer, Berlin, Heidelberg, 2009, pp. 80–86.

[12] J.J. Cabibihan, Design of prosthetic skins with humanlike softness, in: 13th International Conference on Biomedical Engineering, Springer, Berlin, Heidelberg, 2009, pp. 2023–2026, https://doi.org/10.1007/978-3-540-92841-6_504.

[13] J.J. Cabibihan, S.S. Ge, Synthetic finger phalanx with lifelike skin compliance, in: H. Liu, H. Ding, Z. Xiong, X. Zhu (Eds.), Intelligent Robotics and Applications, Springer, Berlin Heidelberg, 2010, pp. 498–504, https://doi.org/10.1007/978-3-642-16587-0_46.

Chapter 5

Artificial skin hysteresis, relaxation, creep, and recovery

5.1 Introduction

Hysteresis refers to the energy dissipated during object contact, as evidenced by the non-coincident loading and unloading curves in the force-displacement characteristic curves [1]. Hysteresis, a characteristic behavior of viscoelastic materials [1], is a serious problem for embedded tactile sensors. If the material was assumed to be linearly elastic but is actually viscoelastic, it is possible that the embedded sensor will remain in a continuously active state even after the contact on the skin surface is removed.

Negligible hysteresis is a desired property for tactile sensing [2][3]. Yet, the literature gives examples wherein tactile sensing devices are built only to discover in the end that the device could not optimally achieve its intended purpose because the selected embedding material was hysteretic [4][5][6]. Without prior materials testing reported, the embedding skin was assumed to be a purely elastic material [7]. Yet in another example, sensors were embedded without considering the time-dependent behavior of the skin material used [8].

A skin material's time-dependent behavior can be known by subjecting it to at least three types of basic tests: cyclic, relaxation, and creep tests [9][1]. Each of these tests can give useful insights for embedding tactile sensors. A cyclic test can show the hysteresis behavior of a viscoelastic material. Hysteresis is the energy dissipated by the material and can be easily shown when the loading and unloading curves do not coincide [1]. The relaxation and creep tests can show the time-dependent return to the new equilibrium state after a disturbance [1].

Viscoelastic materials subjected to a constant strain will relax and a gradual relaxation in stress can be observed when plotted versus time. Obtaining the relaxation time is important for approximating how long it takes for the skin material to stabilize after contact. With a highly viscoelastic skin, signals from an embedded contact sensor may decay slowly until the skin material stabilizes. In a creep test, an applied constant stress will result in an increased strain in a viscoelastic material. After the stress is released, the material will gradually return to its initial state. If the selected material has a long recovery time, the embedded sensor will continue to be in its active state even after the load on the skin surface is removed.

Humanizing Artificial Touch. https://doi.org/10.1016/B978-0-32-395519-5.00015-5

For evaluating the hysteresis, the relaxation or recovery time of embedded tactile sensors, the tests can be performed even without the aid of an FE model of the skin (e.g., [4][5][6]). However, a validated skin model presents a significant advantage because the time-dependent behavior of candidate skin materials can be investigated even before embedding the sensors.

Similar to the organization in the previous chapter, the next sections present the human and artificial skin studies on the time-dependent properties for hysteresis (Sec. 5.2), relaxation (Sec. 5.3), creep and recovery (Sec. 5.4). Sec. 5.5 summarizes the findings for those time-dependent behaviors of artificial skins.

5.2 Skin hysteresis

5.2.1 Human skin hysteresis

Serina et al. [10] investigated the biomechanical behavior of the fingertip pulp under repeated, dynamic compressive loading conditions. Their study showed the viscoelastic and nonlinear force-displacement characteristics of the fingertip pulp under realistic loading scenarios.

5.2.1.1 Methodology

In their experiments, each participant was asked to repeatedly tap their left index finger on a flat plate at controlled tapping rates and fingertip contact angles, while simultaneous measurements of fingertip contact force and pulp displacement were recorded. The tapping trials were conducted at three fingertip contact angles relative to the horizontal plane (0°, 45°, and 90°) and five tapping rates (0.25, 0.5, 1, 2, and 3 Hz). The fingertip angle was adjusted by varying the palm's distance from the force transducer and securing the hand in position.

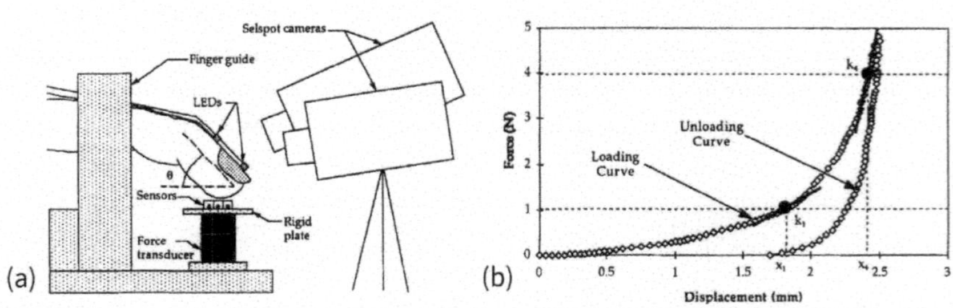

FIGURE 5.1 Repeated compression experiments showing hysteresis. (a) Schematic of the experimental set-up. (b) Force-displacement results for a human fingertip. Reproduced from Serina et al. [10], with permission. ©1997 Elsevier. All rights reserved.

Participants were instructed to maintain the specified fingertip angle and to match the applied force to a positive half sine wave with an amplitude of 5.0 ± 0.5 N, aided by auditory metronome beats and visual feedback of force and reference sine waves. The force was measured by a force transducer, while fingertip displacement was tracked using a two-camera motion analysis system that monitored the three-dimensional motion of two infrared LEDs attached to the fingertip. Contact and release were detected by photoelectric sensors mounted on the plate surface (Fig. 5.1a).

5.2.1.2 Results

The pulp displacement was obtained to be the vertical displacement of the fingertip following contact initiation (Fig. 5.1b). Force-displacement curves were generated for each tap. The pulp stiffness at 1 N and 4 N was calculated as the first derivative of the force-displacement spline. The energy input was calculated as the area under the loading curve. The area between loading and unloading curves is the hysteresis or the energy dissipated.

Results showed that the fingertip pulp exhibited a typical viscoelastic behavior as evidenced by the hysteresis, nonlinear force-displacement curves, and rate-dependence. The hysteresis, represented by the area between the loading and unloading curves, was found to be 2.26 N mm.

For forces below 1 N, the finger pulp was compliant. Even with a low compression force, the finger pulp was compressed to a displacement of around 1.7 mm. With compression forces from 1 N to 4 N, the finger pulp showed a steep increase in slope. From 1.7 mm displacement at 1 N, the displacement at 4 N increased to only around 2.5 mm. At the low-level forces, the mechanoreceptors near the surface are already activated. At the high-level forces, the pulp stiffens rapidly to transfer the forces and impact to the underlying bone structures.

5.2.2 Artificial skin hysteresis

5.2.2.1 Methodology

5.2.2.1.1 Human vs artificial skin hysteresis

The experimental outcomes are significantly impacted by the indentation angle, as demonstrated by Serina et al. [10] in their study. To ensure precise comparisons, it is important that the angles of inclination of the indenter are similar. The majority of experiments conducted on the human fingertip involve indenters that are inclined at an angle ranging from 20° to 40°, as reported in various studies [11][12][13][14].

5.2.2.1.2 Cyclic loading

The fingertip model shown in Fig. 5.2a was used for the simulations on hysteresis. A flat surface was indented into the model with a 0.5 Hz cyclic load applied for 40 s using the positive half of a sine wave. This procedure was similar to the finger tapping experiments described in [10]. The amplitude was set to 1.15 mm for silicone and 0.75 mm for polyurethane. These loading displacements were equivalent to an initial amplitude of 5 N for an artificial finger model with a length of 10 mm. The loading and unloading curves corresponding to the first, tenth and twentieth cycles were plotted. The calculation of hysteresis involved determining the region enclosed by the loading and unloading curves.

In the context of the compression loading of the fingertip pulp, hysteresis can be defined as the mechanical energy dissipated by the skin tissue during a complete cycle of loading and unloading. The hysteresis loss can be calculated as:

$$E_{hyst} = \oint F\, d\delta = \int_0^{\delta_{\max}} [F_{\text{load}}(\delta) - F_{\text{unload}}(\delta)]\, d\delta \tag{5.1}$$

where E_{hyst} is the closed line integral over the loading and unloading curves obtained from the indentation force, F, and the finger pulp displacement, δ. The total hysteresis energy represents the area enclosed between the loading and unloading paths.

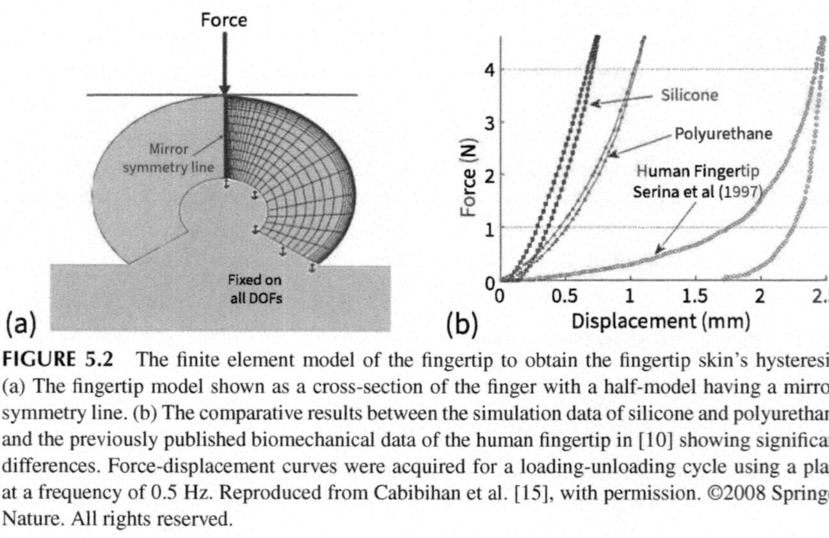

FIGURE 5.2 The finite element model of the fingertip to obtain the fingertip skin's hysteresis. (a) The fingertip model shown as a cross-section of the finger with a half-model having a mirror-symmetry line. (b) The comparative results between the simulation data of silicone and polyurethane and the previously published biomechanical data of the human fingertip in [10] showing significant differences. Force-displacement curves were acquired for a loading-unloading cycle using a plate at a frequency of 0.5 Hz. Reproduced from Cabibihan et al. [15], with permission.

5.2.2.2 Results

5.2.2.2.1 Human vs artificial skin hysteresis

From the results shown in Fig. 5.2b, it can be observed that the synthetic materials exhibited lower levels of compliance in comparison to the human fingertip. The application of a force of 1 N by a plate caused a deformation of 0.28 mm and 0.43 mm on the artificial skin of the fingertip made of polyurethane and silicone, respectively. According to [10], the displacement of the human fingertip was 1.83 mm. The investigation determined that the synthetic materials exhibited minimal hysteresis, with values of 0.17 N mm and

0.29 N mm for silicone and polyurethane, respectively. In contrast, the human fingertip demonstrated a significantly higher hysteresis value of 2.26 N mm [10]. The hysteresis loops were calculated as the area enclosed by the loading and unloading curves (Eq. (5.1)).

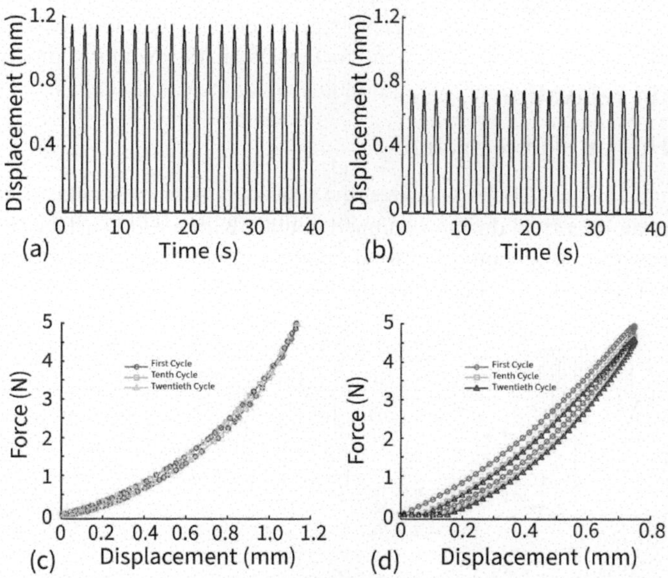

FIGURE 5.3 Skin hysteresis results. Displacement-controlled cyclic loading for (a) silicone and (b) polyurethane, corresponding to 0.5 Hz. The displacement's amplitude is equivalent to 5 N initial force. The hysteresis loops for (c) silicone and (d) polyurethane.

5.2.2.2.2 Hysteresis from cyclic loading

The 0.5 Hz cyclic load to the indenting plate is shown in Fig. 5.3. The amplitude was set to 1.15 mm for silicone (Fig. 5.3a) and 0.75 mm for polyurethane (Fig. 5.3b). This cyclic load condition resulted in the loading and unloading curves of silicone (Fig. 5.3c). The force-displacement curves were consistent from the first cycle until the twentieth cycle. In contrast to silicone, larger hysteresis loops were exhibited by polyurethane in Fig. 5.3d. The computed area at the first cycle has a 22% difference from the tenth cycle (Table 5.1), including a reduction in the magnitude of the peak force. From the tenth cycle, the area diminished to only 3% at the twentieth cycle.

Samples of biological tissues behave with a similar decrease in the hysteresis loops over succeeding cycles before reaching a steady state; a process of initial adjustment known as preconditioning in biomechanical testing [16]. However, such a hysteretic response is contrary to what is commonly expected for practical tactile sensing applications.

TABLE 5.1 Computed Hysteresis Loops (N mm).		
Cycle	**Silicone**	**Polyurethane**
First	0.1723	0.3670
Tenth	0.1717	0.2945
Twentieth	0.1721	0.2854

5.3 Skin relaxation

5.3.1 Human skin relaxation

Pawluk and Howe [12] characterized the time-dependent viscoelastic response of the human fingerpad during sustained contact with a rigid surface (Fig. 5.4). They measured how both the total contact force and the local pressure distribution relaxed over time when the fingerpad was indented.

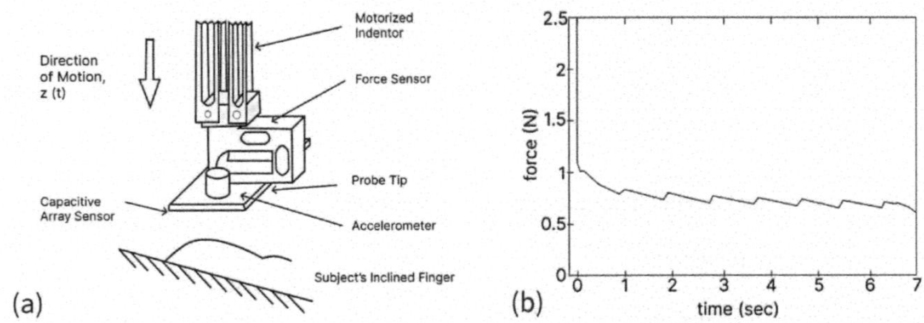

FIGURE 5.4 Experiments on the dynamic contact of the human fingerpad. (a) Schematic diagram of the experiment set-up. (b) Force relaxation results from a participant. Reproduced from Pawluk and Howe [12], with permission.

To model this behavior, they implemented the quasilinear viscoelastic (QLV) model [16] to describe the time-dependent, nonlinear relaxation behavior of the fingerpad. The QLV model has been found to be most applicable for biological tissues. In this model, the material's elastic response is separated from the time-dependent viscoelastic response. To do so, the stress can be expressed as a convolution of the nonlinear elastic function and a reduced relaxation function.

In the QLV model, stress is obtained as:

$$\sigma(t) = \sigma^e(\epsilon(0^+))G(t) + \int_0^t G(t-\tau)\frac{d\sigma^e(\epsilon(\tau))}{d\epsilon(\tau)}\frac{d\epsilon(\tau)}{d\tau}d\tau \qquad (5.2)$$

where σ^e is the instantaneous elastic response (with the superscript e as the purely elastic behavior prior to any viscous relaxation), $\epsilon(0^+)$ is the initial step strain evaluated infinitesimally after time zero (denoted by 0^+ to show the state immediately following the instantaneous loading), $\epsilon(\tau)$ is the strain history, t is the current observation time, and τ is the historical time variable of integration.

The reduced relaxation function, $G(t)$ can be written in the form of a Prony series as below [17]:

$$G(t) = G_\infty + \sum_{i=1}^{N} G_i e^{\left(-\frac{t}{\tau_i}\right)} \tag{5.3}$$

where G_∞ is the long-term (equilibrium) modulus fraction, G_i are the modulus coefficients, and τ_i are the corresponding relaxation times (subject to the normalization constraint $G_\infty + \sum_{i=1}^{N} G_i = 1$).

5.3.2 Artificial skin relaxation

5.3.2.1 Simulation procedure

Three fingertip models with 1, 3 and 5 mm thickness were simulated. A plate was indented into the fingertip models with a 20% nominal strain. This corresponds to 0.2, 0.6 and 1 mm indentations for the 1, 3 and 5 mm skin thicknesses, respectively. The loading rate was set to 1 mm/s and was kept constant until the next 100 s. Fig. 5.5a shows the axisymmetric model of the 3 mm thick skin. The measured reaction force from the plate was obtained. A node at the "bone" surface was selected as shown in the figure. The nodal force and pressure for each of the skin models were also obtained. These correspond to the vertical reaction force (RF2) and mean pressure (PRESS) in Abaqus, respectively. The relaxation time was estimated to be the time when the nodal pressure has "leveled-off", similar to the relaxation criterion used in [16].

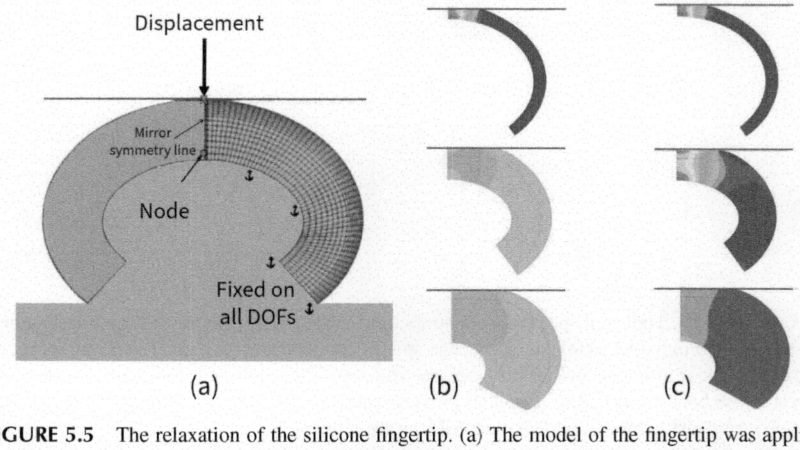

FIGURE 5.5 The relaxation of the silicone fingertip. (a) The model of the fingertip was applied with a displacement-controlled indentation equivalent to 20% nominal strains of the 1, 3 and 5 mm thick skins. The contours of the mean pressure for the 1, 3 and 5 mm thick skins, respectively for (b) silicone and (c) polyurethane samples. These contours were taken at time equals 100 s.

5.3.2.2 Results

With an applied constant displacement, the reaction forces on the plate and on the node for the silicone models exhibited immediate relaxation from about 1 s after the displacement was held static. The magnitudes of the nodal forces were filtered representations of the plate force. Fig. 5.5b depicts the contours of the mean pressure taken at T = 100 s for the different skin thicknesses.

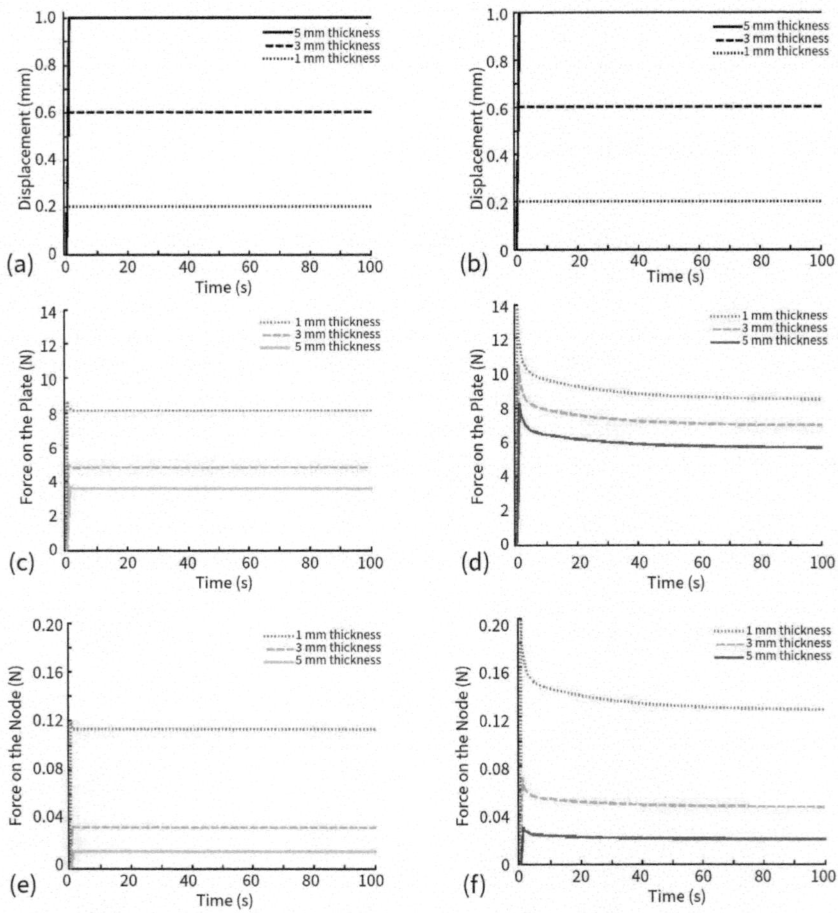

FIGURE 5.6 The applied displacement-controlled indentation equivalent to 20% nominal strains of the 1, 3 and 5 mm thick skins for (a) silicone and (b) polyurethane samples. The resulting forces on the plate for (c) silicone and (d) polyurethane samples. The resulting forces taken on the node nearest to the bone for (e) silicone and (f) polyurethane samples. This location approximates the location where tactile sensors will be mounted over the artificial skin.

Figs. 5.6c-5.6f show the nodal force over time. Similar to the results of silicone, the profiles of polyurethane's nodal reaction forces were a filtered response of the force plate. However, the relaxation to stable levels was longer as compared to silicone. From Fig. 5.6f, the level-off time for the polyurethane models can be seen to go beyond 60 s. Had force or pressure sensors been embedded near the location of the node, it is predicted that the sensor signals will stabilize only after 60 s.

The relaxation effects of the same polyurethane material that was characterized and analyzed in this work can be qualitatively confirmed by the results of an embedded microsensor on a planar skin of polyurethane (cf. Fig. 9 in [8]). In addition, the problem of drift in the output of a commercially-available robotic skin was attributed to the "mechanical properties of the cover material", which requires at least 20 s of loading to achieve precise pressure measurements [5].

5.4 Skin creep and recovery

5.4.1 Human skin creep and recovery

The human skin exhibits distinct time-dependent mechanical behavior under sustained and released loads, primarily characterized by creep and recovery. These phenomena arise from the skin's viscoelastic nature, which integrates both elastic and viscous responses. When a constant step stress is applied at time zero, the gradual increase in strain is what is referred to as creep. This is characterized by the creep compliance function, $J(t)$, where the strain increases with time under constant stress. When the applied stress is removed, the strain response is referred to as recovery. This is when the material recovers to its original state.

The creep behavior of skin is often modeled using a generalized Kelvin–Voigt or Prony series viscoelastic framework. Under a constant uniaxial stress σ_0, the strain $\varepsilon(t)$ can be expressed using the creep compliance function $J(t)$ as described in Tschoegl [18]:

$$\varepsilon(t) = \sigma_0 \cdot J(t) \tag{5.4}$$

For modeling biological skin, a common form of the creep compliance function incorporates a combination of elastic, viscoelastic, and viscous flow components [17]:

$$J(t) = \frac{1}{E_0} + \sum_{i=1}^{N} \frac{1}{E_i}\left(1 - e^{-t/\tau_i}\right) + \frac{t}{\eta} \tag{5.5}$$

where E_0 is the instantaneous elastic modulus, E_i and τ_i are the moduli and relaxation times associated with the i-th viscoelastic element (Prony term), and η represents a dashpot coefficient capturing long-term viscous flow. This equation reflects how skin deforms rapidly at first due to its elastic components, then slowly over time due to internal fluid flow and structural relaxation.

After the stress is removed at a given time $t = t_r$, the skin begins to recover. The strain during recovery, which is denoted by $\varepsilon_{rec}(t)$, is typically modeled by

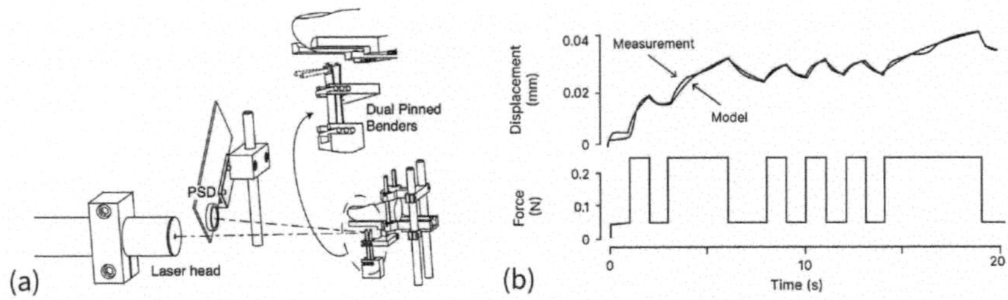

FIGURE 5.7 Experiments on the human fingerpad to show the skin's creeping and recovery behavior. (a) Schematic diagram of the experimental set-up. (b) Creeping and recovery behavior of the skin. Reproduced from Wang and Hayward [20], with permission.

subtracting the time-shifted creep compliance from the peak strain [19]:

$$\varepsilon_{\text{rec}}(t) = \sigma_0 \left[J(t) - J(t - t_r) \right] \tag{5.6}$$

where $J(t)$ is the creep compliance function, which characterizes the time-dependent strain response to a unit step stress in the material, $J(t_r)$ represents the total creep compliance accumulated during the loading period t_r, and $J(t - t_r)$ represents the compliance during the recovery period after the removal of stress.

Wang and Hayward [20] conducted creep experiments on the human fingerpad. Their experiments consisted of applying controlled displacements to a finger pressed against a rigid, smooth surface. The normal and tangential forces and skin deformation were measured with high spatial and temporal resolution (Fig. 5.7a). When the contact force from the stimulus was kept steady, the response showed that the fingerpad skin behaved like a viscoelastic solid and exhibited both elastic recovery and viscous damping (Fig. 5.7b).

5.4.2 Artificial skin creep and recovery

5.4.2.1 Simulation procedure

A creep test requires that a constant force is applied to the material in order to demonstrate the creeping process through displacement or strain measures. An applied concentrated force to a plate must have a balanced distribution of forces to prevent the plate's rotation. This was not possible with the more computationally efficient axisymmetric model used in the previous simulations. Hence, full fingertip models with 1, 3, and 5 mm thickness were used to investigate the skin recovery. Additional simulation conditions represented on the 3 mm skin are shown in Fig. 5.8a. A concentrated vertical force of 10 N was applied through the reference point in the middle of the plate.

The simulation was set such that the time to reach the peak force was kept to 1 s. Then, the force was maintained constant for 5 s. Lastly, the duration of unloading was set to 1 s. Data on the nodal vertical displacement (U2) and

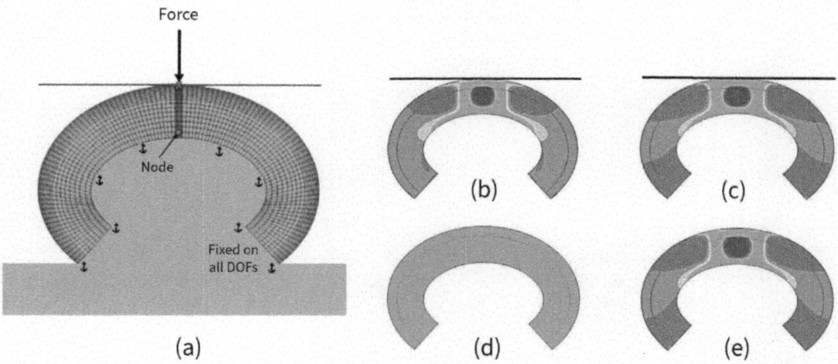

FIGURE 5.8 The finite element model and the resulting contours for silicone and polyurethane samples. (a) The axisymmetric model of the synthetic fingertip in two-dimensional finite element analysis. The shape was designed to resemble a cross-section of the human fingertip. The flat plate indenter was positioned as illustrated. (b) and (c) The contours for the vertical logarithmic strain are shown for the 3 mm skin thickness model for silicone and polyurethane, respectively, at 7 s. (d) and (e) Similar strain conditions for the two sample materials at 17 s for silicone and polyurethane, respectively. Reproduced from Cabibihan et al. [21], with permission. ©2016 Springer Nature. All rights reserved.

vertical logarithmic strain (LE22) were collected from two nodes on the fingertip model.

The first was the node directly below the plate where the concentrated force was applied. The second was the node from an element just above the 'bone' surface as shown in Fig. 5.8a. The recovery time was obtained from the instant the load was released at $T = 7$ s until $T = 17$ s. For each fingertip thickness, the cut-off criterion was set to be at 10% of the strains at the instant when the load was released. This criterion is similar to the one used in [22].

5.4.2.2 *Results*

The strain contours are shown in Fig. 5.8b-5.8e for the silicone and polyurethane material samples. Plotted on Fig. 5.9a for silicone and Fig. 5.9b for polyurethane are the displacements of a node located on the skin surface. Plotted on Fig. 5.9c for silicone and Fig. 5.9d for polyurethane are the displacements of a node near the bone surface. The displacement profile of the subsurface node resembles the profile at the skin surface albeit with smaller magnitude.

The nodal strains for the three skin thicknesses are plotted in Fig. 5.10 for both the silicone and polyurethane. For the magnified results for silicone in Fig. 5.10a at $T = 7$ s, notice that silicone can achieve full recovery within 0.5 s after the release of load from the indenting plate.

For the polyurethane material, the typical behavior of creep in viscoelastic materials is evident from the continued increase of displacement, even as the force is kept constant from $T = 1$ s to $T = 6$ s (Fig. 5.9b). For the three thickness

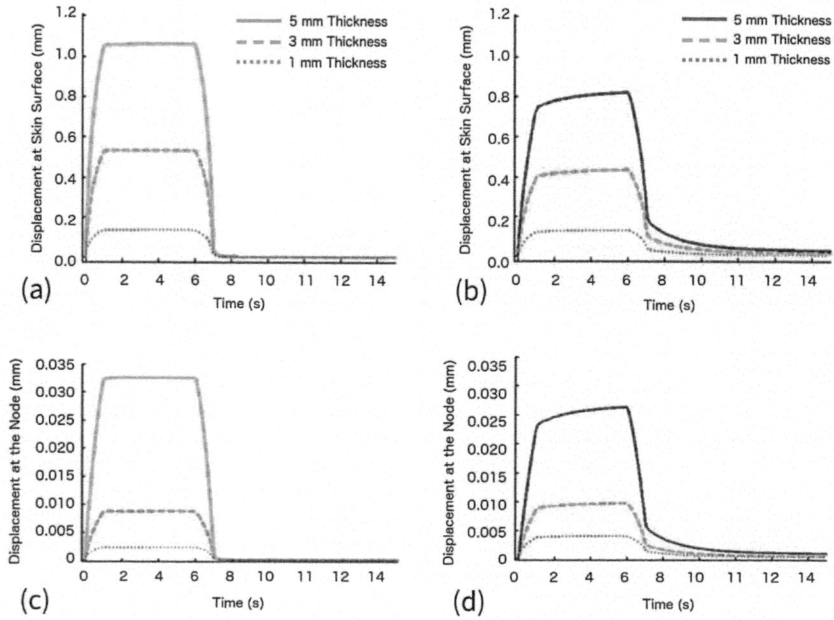

FIGURE 5.9 The creep and recovery of the silicone and polyurethane materials after 10 N constant force was applied for all the skin thicknesses (i.e., 5, 3, 1 mm). The resulting displacements on a node on the skin surface directly below the plate for (a) silicone and (b) polyurethane. To approximate the location of an embedded tactile sensor, the displacement is shown on a node from an element just above the "bone" surface for (c) silicone and (d) polyurethane. In both locations in the skin surface and at the bone surface, the polyurethane sample exhibited creep behavior when a steady load was applied and until the release of the load. The silicone material did not exhibit noticeable creep behavior. Reproduced from Cabibihan et al. [21], with permission.

samples with polyurethane material, Fig. 5.10b suggests that the residual strains from a 10 N load applied for 5 s will be about 0.5% after 10 s of load removal.

5.5 Summary

Understanding the time-dependent behavior of artificial skin materials allows us to better interpret the results coming from the embedded tactile sensors. Consider that even with sensors that have high performance and specifications, the sensor's response will still be dependent on the skin's behavior once it is embedded within the artificial skin.

Hysteresis refers to the energy dissipated during object contact, as evidenced by the non-coincident loading and unloading curves in the force-displacement characteristic curves. In Serina et al. [10], their experimental results on the human fingertip showed that the finger pulp exhibits significant hysteresis. In view of energy dissipation and impact absorption, this attribute is advantageous

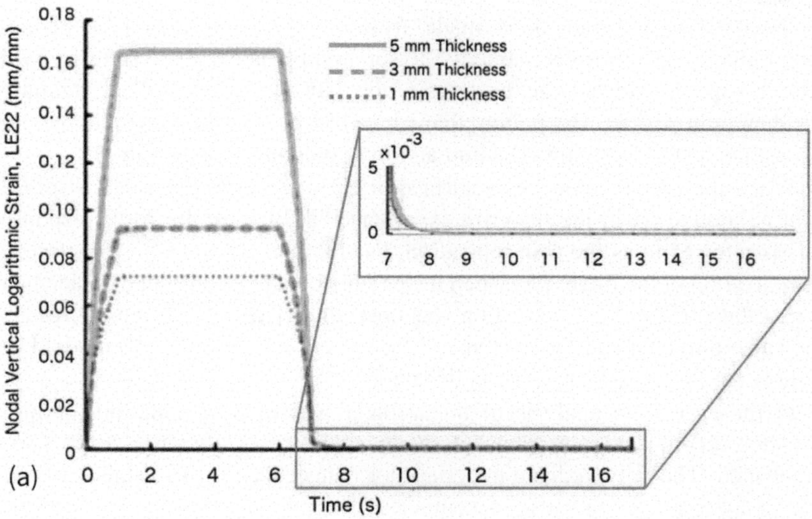

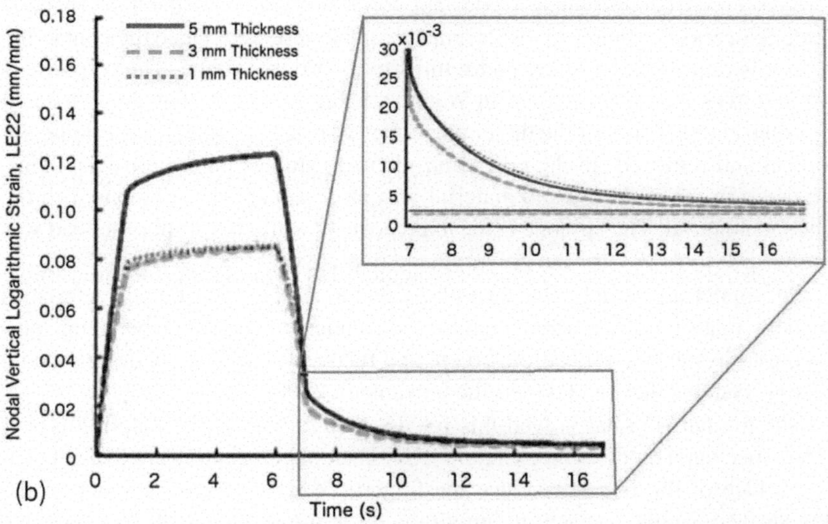

FIGURE 5.10 The nodal logarithmic strains plotted over time for (a) silicone and (b) polyurethane. The insets show the magnified strains immediately after the release of the load at time equals 7 s and at time equals 17 s. Between the two materials, polyurethane took around 10 s to recover from the applied load while silicone took about 0.5 s to recover. Reproduced from Cabibihan et al. [21], with permission. ©2016 Springer Nature. All rights reserved.

for the fingers and the hands. For tactile sensing, the human tactile system has figured out a way to determine that there is no surface or object that is in contact with the finger pulp while it is returning to its initial shape.

In contrast, artificial skins made from materials such as silicone and polyurethane demonstrate markedly lower compliance and hysteresis, with polyurethane showing a larger hysteresis loop than silicone but still substantially less than human skin. The polyurethane has 0.37 N mm while the silicone has half as much at 0.17 N mm. The human finger pulp has 2.26 N mm.

When the fingertip touches a surface or an object and a constant displacement is applied on it, the fingertip experiences a decay in the force response. This is referred to as the skin relaxation. Pawluk and Howe [12] characterized the relaxation of the finger pad when it was subjected to a constant displacement from a flat surface. Skin relaxation was observed in their experimental results. They also implemented Fung's Quasi-Linear Viscoelasticity (QLV) model [16] to show the finger pad's relaxation behavior.

In the relaxation analysis of the artificial skins, the simulations showed that the silicone material achieved steady-state at 1 s after the contact was established. The polyurethane material took longer than 60 s to stabilize. This time delay to relax can cause inaccuracies of the tactile sensing system as the sensor readings take time to stabilize (e.g., [8]).

Skin creep occurs when a constant force is applied to the fingerpad. A gradual increase in fingertip displacement can be observed. Once the contacting surface is removed, the finger pad returns to its original shape. The response is referred to as the skin recovery. In Wang and Hayward [20], they demonstrated the creep and recovery of the finger pulp when a surface contact was periodically applied and removed. In the creep and recovery studies on the artificial skins, the creep in the polyurethane material was more apparent as compared to the silicone material. The silicone achieved recovery within 0.5 s after the load was removed. It took 10 s for the polyurethane to recover.

Soft materials do not behave similarly due to the viscoelastic characteristics of these materials. Knowing their time-dependent mechanical behavior, pertaining to hysteresis, relaxation, creep, and recovery, can help us design tactile sensing systems that can address the unwanted effects.

The human skin has solved this for us through the slowly adapting type 1 (SA I) mechanoreceptors (see Chap. 7). SA I mechanoreceptors are located near the surface of the fingertip. Once the fingertip does not sense contact on the skin surface, signals are sent to ignore the decaying signal. Artificial skins that lack a similar bio-inspired compensation solution will produce prolonged sensor activation even after the contacting surface has been removed. This would make the tactile sensing system useless.

References

[1] G. Holzapfel, Nonlinear Solid Mechanics, A Continuum Approach for Engineering, John Wiley & Sons, 2000, second print.

[2] L.D. Harmon, Automated tactile sensing, The International Journal of Robotics Research 1 (1982 Jun) 3–32, https://doi.org/10.1177/027836498200100201.

[3] E. Kolesar, C. Dyson, Object imaging with a piezoelectric robotic tactile sensor, Journal of Microelectromechanical Systems 4 (1995 Jun) 87–96, https://doi.org/10.1109/84.388117.
[4] B. Gray, R. Fearing, A surface micromachined microtactile sensor array, in: Proceedings of IEEE International Conference on Robotics and Automation, vol. 1, 1996, pp. 1–6, https://doi.org/10.1109/robot.1996.503564.
[5] O. Kerpa, K. Weiss, H. Worn, Development of a flexible tactile sensor system for a humanoid robot, in: Proceedings 2003 IEEE/RSJ International Conference on Intelligent Robots and Systems (IROS 2003) (Cat. No. 03CH37453), vol. 1, 2003, pp. 1–6, https://doi.org/10.1109/iros.2003.1250596.
[6] P. Valdastri, K. Harada, A. Menciassi, L. Beccai, C. Stefanini, M. Fujie, P. Dario, Integration of a miniaturised triaxial force sensor in a minimally invasive surgical tool, IEEE Transactions on Biomedical Engineering 53 (2006) 2397–2400, https://doi.org/10.1109/tbme.2006.883618.
[7] P. Valdastri, K. Houston, A. Menciassi, P. Dario, A. Sieber, M. Yanagihara, Fujie M. Miniaturized, Cutting tool with triaxial force sensing capabilities for minimally invasive surgery, Journal of Medical Devices 1 (2007) 206–211, https://doi.org/10.1115/1.2778700.
[8] L. Beccai, S. Roccella, L. Ascari, P. Valdastri, A. Sieber, M. Carrozza, P. Dario, Experimental analysis of a soft compliant tactile microsensor to be integrated in an anthropomorphic artificial hand, in: Proceedings of 8th Biennial ASME Conference on Engineering Systems Design and Analysis, ESDA2006, 2006, https://doi.org/10.1115/esda2006-95543.
[9] W. Findley, J. Lai, K. Onaran, Creep and Relaxation of Nonlinear Viscoelastic Materials, Dover Publications, Inc., 1993.
[10] E.R. Serina, C. Mote, D. Rempel, Force response of the fingertip pulp to repeated compression. Effects of loading rate, loading angle and anthropometry, Journal of Biomechanics 30 (1997 Oct) 1035–1040, https://doi.org/10.1016/s0021-9290(97)00065-1.
[11] F. Vega-Bermudez, K.O. Johnson, Fingertip skin conformance accounts, in part, for differences in tactile spatial acuity in young subjects, but not for the decline in spatial acuity with aging, Perception & Psychophysics 66 (2004 Jan) 60–67, https://doi.org/10.3758/bf03194861.
[12] D.T.V. Pawluk, R.D. Howe, Dynamic lumped element response of the human fingerpad, Journal of Biomechanical Engineering 121 (1999 Apr) 178–183, https://doi.org/10.1115/1.2835100.
[13] J. Wu, R. Dong, W. Smutz, S. Rakheja, Dynamic interaction between a fingerpad and a flat surface: experiments and analysis, Medical Engineering & Physics 25 (2003 Jun) 397–406, https://doi.org/10.1016/s1350-4533(03)00035-3.
[14] R.S. Johansson, I. Birznieks, First spikes in ensembles of human tactile afferents code complex spatial fingertip events, Nature Neuroscience 7 (2004 Jan) 170–177, https://doi.org/10.1038/nn1177.
[15] J.J. Cabibihan, S. Pattofatto, M. Jomaa, A. Benallal, M.C. Carrozza, Towards humanlike social touch for sociable robotics and prosthetics: comparisons on the compliance, conformance and hysteresis of synthetic and human fingertip skins, International Journal of Social Robotics 1 (2009) 29–40, https://doi.org/10.1007/s12369-008-0008-9.
[16] Y. Fung, Biomechanics: Mechanical Properties of Living Tissue, Springer-Verlag, 1993.
[17] H. Joodaki, M.B. Panzer, Skin mechanical properties and modeling: a review, Proceedings of the Institution of Mechanical Engineers. Part H, Journal of Engineering in Medicine 232 (2018 Mar) 323–343, https://doi.org/10.1177/0954411918759801.
[18] N.W. Tschoegl, The Phenomenological Theory of Linear Viscoelastic Behavior, Springer, Berlin, Heidelberg, 1989.
[19] J.D. Ferry, Viscoelastic Properties of Polymers, 3rd ed., John Wiley & Sons, Nashville, TN, 1980 Sep.
[20] Q. Wang, V. Hayward, In vivo biomechanics of the fingerpad skin under local tangential traction, Journal of Biomechanics 40 (2007) 851–860, https://doi.org/10.1016/j.jbiomech.2006.03.004.

[21] J.J. Cabibihan, M.K. Abu Basha, K. Sadasivuni, Recovery behavior of artificial skin materials after object contact, in: A. Agah, J.J. Cabibihan, A.M. Howard, M.A. Salichs, He H. Cham (Eds.), Social Robotics, Springer International Publishing, 2016, pp. 449–457, https://doi.org/10.1007/978-3-319-47437-3_44.

[22] E. Sladek, R. Fearing, The dynamic response of a tactile sensor, Proceedings - IEEE International Conference on Robotics and Automation 2 (1990) 962–967, https://doi.org/10.1109/robot.1990.126116.

Chapter 6

Artificial skin thickness and filtering

6.1 Introduction

In embedding the tactile sensors underneath soft artificial skin materials, how thick should the artificial skin be? The thickness of the artificial skin has an influence on the pressure transmission from the surface to the embedded tactile sensors. It also has an effect on the ability of an array of embedded sensors to discriminate an externally applied shape from another. By examining these factors, we can gain insights into optimizing the thickness of the skin to ensure optimal balance between sensitivity and protection of the sensors against damage.

The human fingertip skin has a multilayer structure consisting of the epidermis, dermis, and subcutaneous tissues. Each layer has its unique properties for compliance, conformance, and impact absorption. The fingertip layers exhibit remarkable spatial and temporal filtering properties. These layers enable us to discriminate textures, shapes, and other surface features. Such capabilities are attractive for artificial tactile sensing systems to be replicated.

Selecting the thickness of the skin material is important for discriminating the shape that is applied at the skin surface and is detected with the embedded tactile sensors [1][2]. The skin material acts like a low-pass filter [3]. Embedding the sensor with a thick skin will blur the stress or strain signals being transmitted to the sensor. On the other hand, embedding the sensor with a thin skin improves its detection ability but at the same time makes it vulnerable to damage. The balance between robustness and sensitivity is an important consideration for design trade-off decisions, especially for some of the inherently fragile semiconductor-based tactile sensors.

Understanding the effects of artificial skin aids in the solution of the tactile inversion problem. This involves the approaches to decode what is happening on the skin's surface (i.e., the shape or force of an object touching it) using the pressure profiles that are detected by the embedded sensors beneath the artificial skin. Three approaches have been proposed to analyze this problem (i.e., closed-form solutions, numerical integration, and finite element modeling) and are presented in this chapter.

The next sections are organized according to the thickness effects of human skins and then artificial skins. Sec. 6.2 describes the spatial and filtering effects

Humanizing Artificial Touch. https://doi.org/10.1016/B978-0-32-395519-5.00016-7

of the human skin's thickness. In Sec. 6.3, I present the various approaches for detecting the external shape through subsurface sensors, which can then give insights on the appropriate thickness to embed the tactile sensors. Lastly, Sec. 6.4 gives the summary for this chapter.

6.2 Effects of human skin thickness

The human fingertip skin acts as a filter in the transmission of tactile signals to the mechanoreceptors [4][5]. The finger pad exhibits a nonlinear viscoelastic characteristic, which makes it act as a temporal and spatial filter for the transmitted tactile signals [6]. The skin layers (i.e., the epidermis, dermis, and subcutaneous tissue) modulate stimuli before they reach the mechanoreceptors. The skin converts mechanical signals into electrical signals, and in the process, enhances or dampens the mechanical stimuli before they reach the sensors. Scheibert et al. [4] and Prevost et al. [7] showed that the skin ridge geometry and orientation modulate vibration spectra when the fingerpad was presented with fine textures with spatial scales under 200 μm.

Vallbo and Johansson [8] and Phillips et al. [9] showed that the skin modulates the waveform and energy of tactile inputs before these signals reach mechanoreceptive afferents. As a result, activation thresholds and response patterns differ across receptor types. The somatosensory system then uses the filtered responses to discriminate fine surface properties, object edges, and textures under various conditions.

6.2.1 Thickness of the human fingertip skin

The glabrous skin of the fingertip is organized in distinct layers with characteristic thicknesses that influence mechanoreceptor placement and skin mechanics. When measured from the surface of the fingertip, Kandel et al. [10] illustrate the epidermis to be approximately 1 mm thick, which is composed of a topmost layer of stratum corneum with a 0.1 to 0.7 mm thickness and the rest is the epidermis tissue. Gerling et al. [11] presented an FEA model of the fingertip with the epidermis thickness set to be 0.471 mm, with the stratum corneum to be 0.371 mm and the epidermis to be 0.1 mm. The dermis was modeled to have a 1.153 mm thickness.

6.2.2 Temporal filtering

The viscoelastic nature of the fingertip skin causes it to act as a temporal filter. Upon contact with an object or a surface, the transient forces and vibrations are attenuated based on their frequency and duration. This kind of behavior has been shown in tactile simulations and confirmed through experimental recordings [12][13].

For example, high-frequency vibrations are efficiently transmitted through deeper layers of the skin to activate Pacinian corpuscles, while lower-frequency or spatially localized stimuli tend to activate Meissner corpuscles or Merkel cells in superficial layers [14][15].

6.2.3 Spatial filtering

Spatial filtering is primarily affected by the epidermal ridges (fingerprint patterns) and the shape of the dermal papillae, which modify the distribution of mechanical stress on the skin's surface as well as underneath it [16]. The interested reader is referred to Figs. 2.1, 2.2 and 7.2 for the figures of the external and internal ridges.

The ridged structures on the finger's surface and underneath it allow better edge detection and texture differentiation [17]. Through a mechanical lever effect on these micro-sized ridges [18], pressure is localized toward the mechanoreceptors connected on those ridges to improve tactile sensing.

6.3 Effects of the artificial skin thickness

The artificial skin's thickness has an effect on the pressure readings that will be recorded beneath the artificial skin's surface. Thus, determining the thickness is important to discriminate the surface features that are presented on the skin surface. This information is helpful for the perception of shapes or for proper grasping or manipulation.

In this context, the succeeding sections describe what is known as the tactile inversion problem [19][20]. This problem is further stated as follows: *With the stress and/or strain measured on the subsurface of the artificial skin, decode the contact pressure or deflection on the skin surface* [2][21].

At least three approaches have been proposed to solve the tactile inversion problem. Zhang and Chen [20] summarized these according to closed form, numerical integration, and finite element modeling approaches. This chapter will show a case study on the finite element approach to gain insight on the surface contact from the subsurface pressure profiles.

6.3.1 Closed-form approach

For contact analysis problems, the fingertip can be approximated as a linear elastic half-space [22]. In addition, plane strain assumptions are used. This implies that there is no variation in the surface shape and contact conditions along the entire width of the body. Three representative contact conditions that have closed-form solutions are shown next for point contact, rigid line contact, and rectangular line contact (Fig. 6.1).

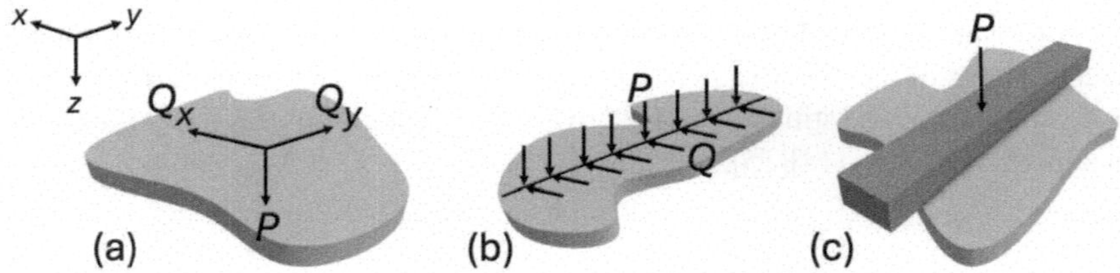

FIGURE 6.1 Representative contact conditions with closed form solutions: (a) point contact, (b) rigid line contact, and (c) rectangular surface contact.

6.3.1.1 *Point contact*

When a point contact is applied (Fig. 6.1a), a normal force, P, with two tangential forces, Q_x and Q_y, can be observed. The subsurface stress and strains can be calculated as below [22][20]:

$$\sigma_z(x, y, z) = -\frac{3z^2}{2\pi r^5}\left(Pz + Q_x x + Q_y y\right) \tag{6.1}$$

$$\varepsilon_z = -\frac{3\left(z^2 - \frac{1}{2}\left(x^2 + y^2\right)\right)}{2\pi r^5 E}\left(Pz + Q_x x + Q_y y\right) \tag{6.2}$$

where $r = (x^2 + y^2 + z^2)^{1/2}$ is the distance from the point of the applied force to the target location. The Young's modulus of the material is denoted as E. The contact state vector is denoted as $S = [Q_x, Q_y, P]^T$.

6.3.1.2 *Rigid line contact*

When a rigid line contact or an edge is applied on a planar elastic surface along the y-axis (Fig. 6.1b), the normal stress and strain along the z are given below [22][20]. These equations further assume that only a tangential force perpendicular to the edge, but not along the edge, exists:

$$\sigma_z = -\frac{2z^2\left(Pz + Qx\right)}{\pi r^4} \tag{6.3}$$

$$\varepsilon_z = \frac{3}{2\pi r^4 E}\left(Pz + Qx\right)\left(z^2 - x^2\right) \tag{6.4}$$

where P is the normal force, Q is the tangential force applied by the line load in force per unit length along the line, and E is the Young's modulus. There are no resulting moments for a simple loading condition.

6.3.1.3 *Rectangular surface contact*

When a normal force is symmetrically applied by a flat rectangular block with a width of $2a$ on a planar elastic surface along the y-axis (Fig. 6.1c), the surface pressure distribution or the surface traction is calculated below [22][20]:

$$p(x) = \begin{cases} -\frac{P}{\pi}\frac{1}{\sqrt{a^2-x^2}}, & -a \le x \le a \\ 0, & \text{otherwise} \end{cases} \tag{6.5}$$

wherein the subsurface stress distribution can be obtained in terms of polar coordinates as:

$$\sigma_z(r,\theta) = -\frac{P}{\pi r_{12}}\left(\cos\theta_{12} + \frac{r^2}{r_{12}^2}\cos\theta\cos(\theta - 3\theta_{12})\right) \tag{6.6}$$

where $r_{12} = (r_1 r_2)^{1/2}$ and $\theta_{12} = \frac{1}{2}(\theta_1 + \theta_2)$. Eq. (6.6) shows the r_1, r_2, θ_1, and θ_2 as functions of a, r, and θ, respectively (see Fig. 6.2 for the variable definitions).

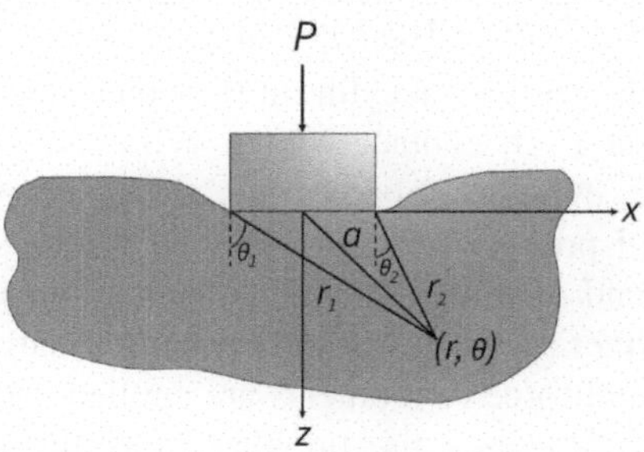

FIGURE 6.2 Rectangular surface contact on an elastic skin with the definitions of r_1, r_2, θ_1 and θ_2.

6.3.2 Numerical integration approach

Numerical integration solutions can be used when closed-form solutions do not exist. Zhang and Chen [20] proposed a solution where the surface traction is derived from the externally applied force and the contacting object's shape and dimensions. The distribution of the subsurface stress can then be obtained by considering the contact as a collection of points or line contacts. Through superposition, the effects of the contacts are integrated by applying the stress equations in Eq. (6.1) and Eq. (6.3).

6.3.2.1 Spherical object on a planar sensor

When the rigid sphere comes into contact with the body, a normal force, P, and a tangential force, Q, are obtained. The area of contact interface results into a circular contact with the radius given by:

$$\phi = \left[\frac{9Pr_s}{16E}\right]^{1/3} \tag{6.7}$$

where the radius of the spherical object is denoted by r_s. The polar coordinates for the surface traction in the normal and tangential coordinates are given by:

$$\sigma_z(r) = \frac{3P(\phi^2 - r^2)^{1/2}}{2\pi\phi^3} \tag{6.8}$$

$$\sigma_x(r) = \frac{3Q(\phi^2 - r^2)^{-(1/2)}}{2\pi\phi} \tag{6.9}$$

where P is the normal force and Q is the tangential component of the force vector. The subsurface stress or strain can be obtained by integrating Eq. (6.8) and Eq. (6.9) with the use of the equations for stress and strain for the point contact in Eq. (6.1) and (6.2).

6.3.2.2 *Cylindrical object on a cylindrical sensor*

A cylindrical robot finger and a cylindrical object can be considered as a cylindrical sensor that is in contact with a cylindrical object. The contact analysis between two cylindrical bodies has earlier been provided [23][22]. Zhang and Chen [20] proposed to solve this by finding the area of contact and surface traction and then numerically integrating the effect of the surface traction. It can be further assumed that only normal force, P, is exerted by the object and no tangential force or moments are applied. The surface traction is given by:

$$p(x, y) = \frac{3P}{2\pi AB}\sqrt{1 - \frac{x^2}{A^2} - \frac{y^2}{B^2}} \tag{6.10}$$

where the contact area results into an ellipse and the resulting major axis, A, and minor axis, B, have dependencies on the object's radii of curvature and the sensor, the material constants of the elastic surface, contact force, and the angle between the axes of the cylindrical bodies [22]. Using Eq. (6.10), the subsurface stress or strain can be calculated using the equations for stress and strain for the point contact in Eq. (6.1) and (6.2).

6.3.3 Finite element modeling approach

Finite Element Modeling is another alternative approach to solve contact problems. With this approach, any user-defined arbitrary object, sensors geometries, and loading conditions can be analyzed. This approach offers benefits when the closed-form approach and the surface traction technique in the numerical integration approach become limiting. With a linear elastic FE model of a planar artificial skin, Shimojo [3] showed that the skin covering blurs the signal from an externally applied load.

6.3.3.1 Fingertip model

Fig. 6.3 shows a model of an artificial finger, which includes the skin, bone, a rigid tissue, and a nail. In order to optimize computational performance, just half of the cross section of the fingertip skin was modeled. A fingertip finite element model was constructed using the Abaqus (Dassault Systemes, Simulia Corp., Providence, RI, USA). The mesh used was an 8-node biquadratic element type for plane strain analysis. A unit of thickness equal to 1 was established, with measurements directed towards the page. The interface located at the lower part of the model, specifically the area consisting of the bone, stiff tissue, and nail, was fixed in all degrees of freedom.

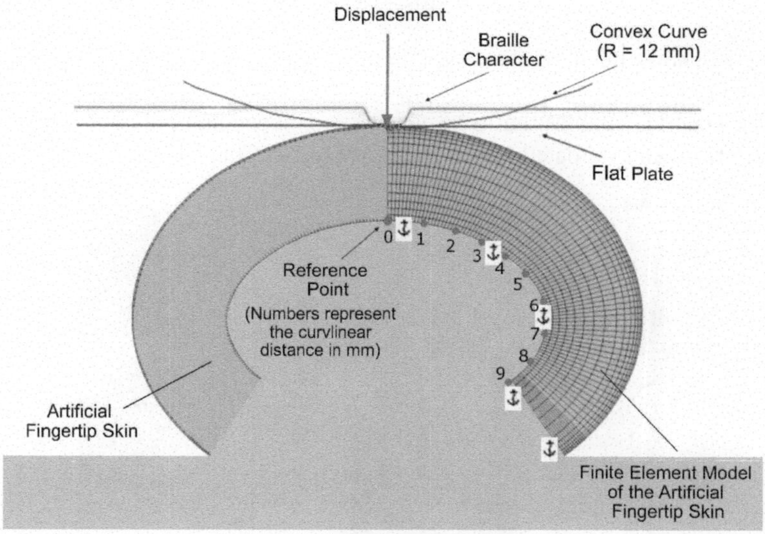

FIGURE 6.3 Artificial fingertip model. The artificial fingertip model positioned over a stiff structure. The structure represents the human bone, stiff finger tissues, and nail. The simulation was performed only on the meshed geometry to optimise computing efficiency. The fingertip model additionally displays the precise positions where the plate, convex curve, and Braille character were applied. The numbers on the surface of the artificial bone indicate the curvilinear distance, measured in millimeters.

The geometry provides an approximation of the rigid bone, nail, and the tissue that exists between them. Subsequently, three fingertip models with thicknesses of 1, 3, and 5 mm were simulated.

The indenters were configured to move vertically downwards towards the fingertip. The indenters featured a flat surface, a convex curve with a radius of 12 mm, and a Braille character's cross-section. The geometry of the Braille character was derived from the technical standards of the Standard American measurements. These criteria recommend that the Braille character should have a height of 0.48 mm and a base diameter of 1.44 mm. The three surface types were subjected to a 20% nominal strain, which corresponded to indentations of 0.2, 0.6, and 1 mm for the fingertip model's thicknesses of 1, 3, and 5 mm, respectively. The indentation rate was 1 mm/s. Measurements were taken for each of the skin thicknesses at the last time interval.

The mean pressure and two components of shear were proposed as the simplest independent sensors for decoding tactile signals [24]. To achieve proper indentation of surface shapes, it is necessary to provide an adequate mean pressure. Therefore, the nodes located on the surface of the artificial bone were chosen, and the starting node for the path was designated as the 0 reference point, as seen in Fig. 6.3. The mean pressure, P, commonly referred to as the mean normal stress, is equal to:

$$P = \frac{1}{3}(\sigma_x + \sigma_y + \sigma_z) \tag{6.11}$$

where σ denotes the normal stresses on those axes, assuming no shear stresses involved [25].

The material samples that were used (i.e., silicone and polyurethane) were characterized and validated earlier and described in Chap. 3.

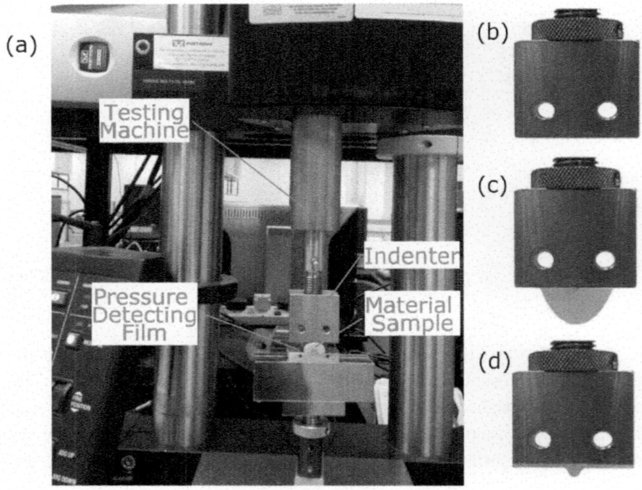

FIGURE 6.4 Experimental set-up. (a) Indentation of the material sample by the indenters. (b) The flat surface indenter. (c) The curved indenter. (d) The Braille surface indenter.

6.3.3.2 *Validation experiments*

The finite element model to determine the subsurface pressure distributions of the artificial skins have been validated experimentally. The detailed experimental setup, the sensor technique used, and testing procedures were earlier described in Cabibihan et al. [2]. In summary, the materials and equipment utilised consisted of Pressurex-micro sensor film for pressure mapping and a universal testing machine (Fig. 6.4a). Specialised indenters were created specif-

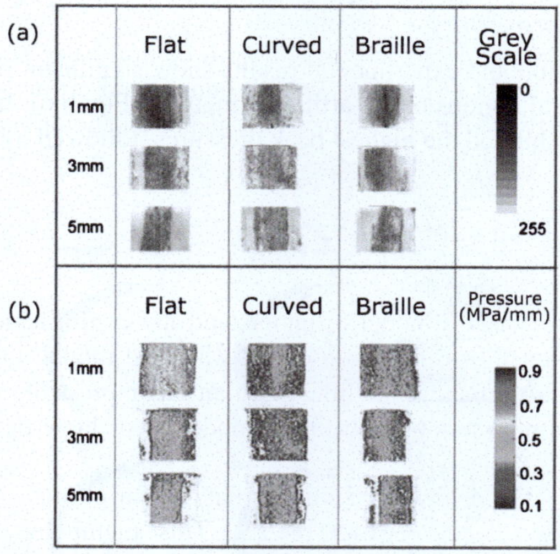

FIGURE 6.5 Pressure film results for the silicone material. (a) Grayscale images obtained from the indentation experiment. (b) Converted color maps to obtain the pressure measurements.

ically for the experiment: flat surface (Fig. 6.4b), curved surface (Fig. 6.4c), and Braille surface with a height of 0.48 mm and a base diameter of 1.44 mm (Fig. 6.4d).

The pressure sensing film underwent calibration by conducting controlled load tests to establish a correlation between pressure and grayscale value. Based on the mean grayscale value, the applied pressure was interpreted using the resulting linear function (with 90% confidence bounds).

The artificial skin materials were subjected to indentation testing using indenters at a nominal strain of 20%. This led to varying indentation depths depending on the thickness of the skin. The indentation rate was 1 mm/s. Fig. 6.5a displays the results in greyscale.

The pressure films, together with the greyscale results, were scanned using a desktop scanner and subsequently analysed using MATLAB®-based software. The software transformed the greyscale numbers into pressure measurements and visualised them as colour maps. The system examined every pixel and generated corresponding pressure readings, which were then visualised using a colour bar (Fig. 6.5b).

6.3.4 Results

Fig. 6.5a shows the experimental results obtained by placing pressure detecting films between the layers of artificial skin and the curved surface of a rigid structure. Fig. 6.5b displays the color maps of the pressure contours obtained. The colour differences were primarily focused on the contact surface. It is possible to think of the set area at the bottom of the fingertip model as a bone structure. This region acted as the sink for the mechanical signals.

6.3.4.1 Peak pressures

Both the simulation and experimental results show significant pressure magnitudes at the central regions of the artificial fingertip (Fig. 6.6). It is evident that the flat surface exhibited the highest peak pressures, followed by the curved and

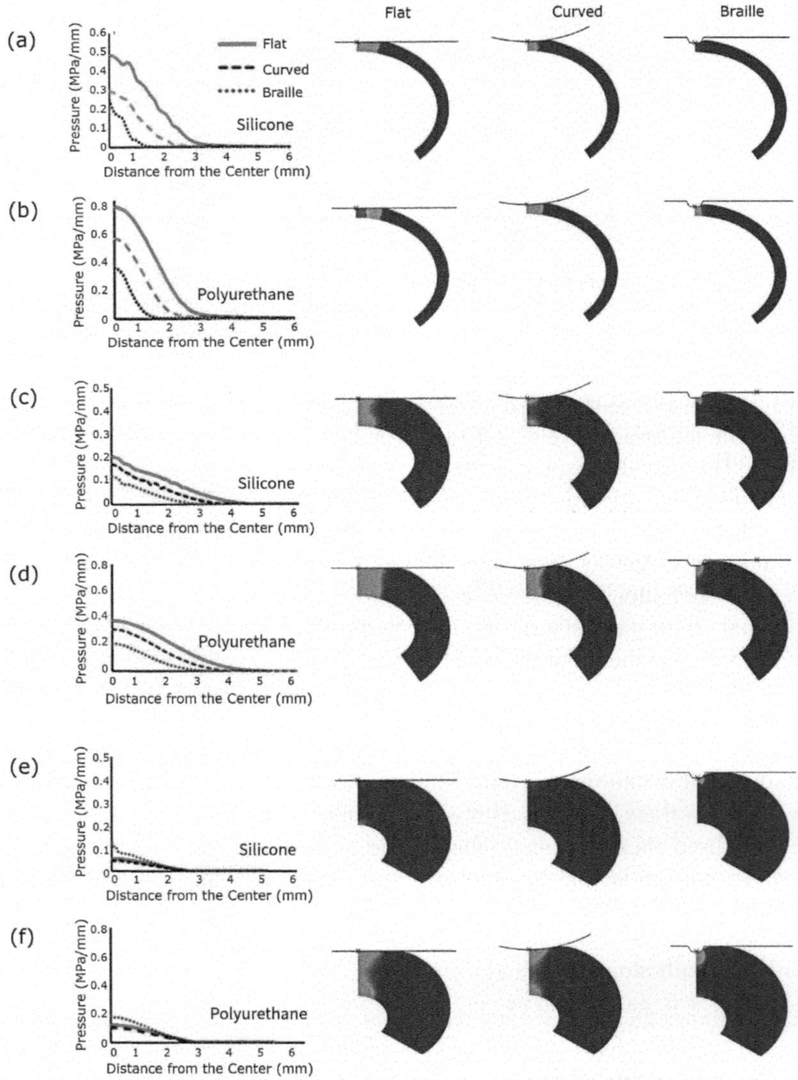

FIGURE 6.6 Pressure contours of 1 mm, 3 mm, and 5 mm artificial skins from validated simulation results. Simulation results for the following artificial skin thickness for silicone and polyurethane, respectively: (a) (b) 1 mm, (c) (d) 3 mm, and (e) (f) 5 mm for flat, curved, and Braille character surfaces, which were indented on the fingertip surfaces.

Braille surfaces for the 1 mm and 3 mm thick skins. However, this trend was not observed with the 5 mm thick skin.

As the thickness of the skin increases, the amplitude of pressure decreases. Put simply, the thickness of the skin reduces the amount of pressure that is transferred to the sensors located below. Thicker skin provides protection for subsurface sensors by absorbing the pressure that reaches them. Consequently, the pressure signals may be sufficiently small to be accurately detected.

6.3.4.2 Pressure profiles' spans

The thickness of the synthetic skin impacts the extent of the pressure dispersion beneath its surface. The pressure spread shifts less widely as the skin gets thicker. In general, the flat indenter has the biggest span for 1 mm and 3 mm thick skins made of silicone and polyurethane. The curved indenter comes in second, and the Braille indenter comes in third (Fig. 6.6).

Regarding the span of the 5 mm thick skins, the three indenter varieties exhibited negligible variation. Due to the thickness of the skin, the profiles of flat, curved, and Braille were already indistinguishable beneath the surface. The distances from the centre of the 5 mm thick skins for both silicone and polyurethane ranged from 2.3 to 2.5 mm, depending on the type of indenter used.

6.3.4.3 Pressure profiles' shapes

The shape of the pressure profiles can also be influenced by the thickness of the synthetic skin. Fig. 6.6 demonstrates that the pressure profiles generated by the plate, curved surface, and Braille character can be distinguished from one another for both silicone and polyurethane materials in the 1 mm and 3 mm thick skins. The 5 mm-thick skins have already overlapped with the profiles of the plate and the curved surface. The Braille character fully penetrated the skin and created an indentation with a height of 0.48 mm. Regardless of the three thicknesses involved, the Braille character was not visible in the silicone or polyurethane results.

In Salehi et al. [16], they used machine learning techniques to further discriminate flat, curved, or wedged indenters. They used a 2×2 element tactile force sensor array and Naïve Bayes (NB), Artificial Neural Networks (ANN), and Support Vector Machines (SVM) algorithms to achieve the discrimination task.

6.4 Summary

Understanding the effect of the skin thickness of an artificial skin gives us the possibility to optimize a skin material's thickness to serve both its sensing task and its practical requirement to be robust. An artificial skin offers the softness, conformance, and the absorption of impact from the skin's contact interaction with the external world. If we desire a general purpose artificial hand that can work in grasping and manipulating objects in daily life as well as to catch a thrown object like a ball, the sensors that provide the feedback to the control system need to do their main job of sensing. They also need to be protected from damage and from the wear and tear in accomplishing their daily tasks.

The chapter first presented information about the thickness of the human fingertip skin. The fingertip skin was shown to have a multilayer organization with topmost layers (i.e., the epidermis and dermis) to be around 2 mm from the surface. The effect of these two layers is not to be ignored. The layers have ridges that help filter signals that are sent to the mechanoreceptors connected to the internal ridges. Both normal and tangential forces can stimulate the mechanoreceptors to determine contact or textures. Below the epidermis and dermis layers are the subcutaneous tissue. This layer has fatty, fluidlike substances that have cushioning properties to absorb impact. Taken together, the fingertip skin's fine structures and viscoelastic properties give it spatial and temporal filtering characteristics to aid in human tactile sensing for gross and fine manipulation and recognition of smooth or rough surfaces.

One of the main challenges in tactile sensing is the so-called tactile inversion problem, where the contact pressure or deflection on the skin surface must be decoded through the embedded sensors that have to sense stress and/or strain. The skin plays a significant role in solving this problem. Three approaches were proposed in the literature. Closed-form solutions [26][27][19] make use of mathematical expressions to solve for the subsurface stress and strains. The contact conditions, such as a point, line, or rectangular surfaces, are idealized. Linear elasticity and plane strain conditions are assumed. The analysis becomes simple but the applications in complex geometries are limited.

The numerical integration approach [28][20][24] extends the closed-form solution approach. Arbitrary contact shapes are approximated as a collection of point or line contacts that permit the superposition of their effects. The finite element modeling [29][21][20] provides the most flexibility among the three approaches. In this technique, arbitrary contact geometries, material properties and loading conditions can be used in the analyses. The validated FEA model in Cabibihan et al. [1] [2] was used in this chapter to give evidence that the skin thickness affects the pressure profiles that are transmitted underneath the skin.

In the simulations, fingertip geometries with thicknesses of 1 mm, 3 mm, and 5 mm, were subjected to indentations from flat, curved, and Braille character indenters. The results imply that the use of near-surface touch sensors has the potential to enhance the ability to discriminate between various shapes. Earlier neuroscientific studies [9][30][31] showed that the SA I and FA I types of mechanoreceptors are known to be sensitive to surface features and orientations of objects.

The results further show that thinner skins (e.g., 1 mm and 3 mm) preserve the unique features of the pressure profiles to allow shape discrimination. For example, peak pressures were highest with flat indenters, followed by curved and Braille ones, but decreased significantly with increasing thickness due to enhanced filtering. At 5 mm thickness, the pressure profile's span narrowed to 2.5 mm from the center. At this thickness, the pressure profiles for different indenters cannot be distinguished from one another.

In previous chapters, the material choice between silicone and polyurethane had a significant effect on their conformance, compliance, and the time-dependent properties related to hysteresis, relaxation, creep, and recovery. In this chapter, the thickness of the skin also plays a significant role in modulating the signals that reach the bottom layers of the skin. The choice of materials and thickness has a direct effect on the sensing system, which can trigger the subsequent event detection algorithms for tightening or loosening a grasp on an object, or for discriminating between a smooth or rough surface.

References

[1] J.J. Cabibihan, M.C. Carrozza, Influence of the Skin Thickness on Tactile Shape Discrimination, 2012 4th IEEE RAS & EMBS International Conference on Biomedical Robotics and Biomechatronics (BioRob), IEEE, 2012, pp. 1681–1685, https://doi.org/10.1109/biorob.2012.6290938.

[2] J.J. Cabibihan, S.S. Chauhan, S. Suresh, Effects of the artificial skin's thickness on the subsurface pressure profiles of flat, curved, and Braille surfaces, IEEE Sensors Journal 14 (2014 Jul) 2118–2128, https://doi.org/10.1109/jsen.2013.2295296.

[3] M. Shimojo, Mechanical filtering effect of elastic cover for tactile sensor, IEEE Transactions on Robotics and Automation 13 (1997) 128–132, https://doi.org/10.1109/70.554353.

[4] J. Scheibert, S. Leurent, A. Prevost, G. Debrégeas, The role of fingerprints in the coding of tactile information probed with a biomimetic sensor, Science 323 (2009 Mar) 1503–1506, https://doi.org/10.1126/science.1166467.

[5] B.M. Dzidek, M.J. Adams, J.W. Andrews, Z. Zhang, S.A. Johnson, Contact mechanics of the human finger pad under compressive loads, Journal of the Royal Society Interface 14 (2017), https://doi.org/10.1098/rsif.2016.0935.

[6] D.T.V. Pawluk, R.D. Howe, Dynamic lumped element response of the human fingerpad, Journal of Biomechanical Engineering 121 (1999 Apr) 178–183, https://doi.org/10.1115/1.2835100.

[7] A. Prevost, J. Scheibert, G. Debrégeas, Effect of fingerprints orientation on skin vibrations during tactile exploration of textured surfaces, Communicative & Integrative Biology 2 (2009) 422–424, https://doi.org/10.4161/cib.2.5.9052.

[8] Å.B. Vallbo, R.S. Johansson, Properties of cutaneous mechanoreceptors in the human hand related to touch sensation, Human Neurobiology 3 (1) (1984) 3–14.

[9] J. Phillips, R. Johansson, K. Johnson, Responses of human mechanoreceptive afferents to embossed dot arrays scanned across fingerpad skin, The Journal of Neuroscience 12 (1992 Mar) 827–839, https://doi.org/10.1523/jneurosci.12-03-00827.1992.

[10] E.R. Kandel, J.H. Schwartz, T.M. Jessell, S.A. Siegelbaum, A.J. Hudspeth, Principles of Neural Science, 5th ed., McGraw-Hill Education/Medical, 2012 Oct.

[11] G.J. Gerling, I.I. Rivest, D.R. Lesniak, J.R. Scanlon, L. Wan, Validating a population model of tactile mechanotransduction of slowly adapting type I afferents at levels of skin mechanics, single-unit response and psychophysics, IEEE Transactions on Haptics 7 (2014 Apr) 216–228, https://doi.org/10.1109/toh.2013.36.

[12] M. Srinivasan, Surface deflection of primate fingertip under line load, Journal of Biomechanics 22 (1989 Jan) 343–349, https://doi.org/10.1016/0021-9290(89)90048-1.

[13] J.Z. Wu, R.G. Dong, Analysis of the contact interactions between fingertips and objects with different surface curvatures, Proceedings of the Institution of Mechanical Engineers.

Part H, Journal of Engineering in Medicine 219 (2005) 89–103, https://doi.org/10.1243/095441105x9327.
[14] H.P. Saal, S.J. Bensmaia, Touch is a team effort: interplay of submodalities in cutaneous sensibility, Trends in Neurosciences 37 (2014 Dec) 689–697, https://doi.org/10.1016/j.tins.2014.08.012.
[15] B. Delhaye, A. Barrea, B.B. Edin, P. Lefèvre, J.L. Thonnard, Surface strain measurements of fingertip skin under shearing, Journal of the Royal Society Interface 13 (2016 Feb) 20150874, https://doi.org/10.1098/rsif.2015.0874.
[16] S. Salehi, J.J. Cabibihan, S.S. Ge, Artificial skin ridges enhance local tactile shape discrimination, Sensors 11 (2011) 8626–8642, https://doi.org/10.3390/s110908626.
[17] J.R. Phillips, K.O. Johnson, Tactile spatial resolution. II. Neural representation of bars, edges, and gratings in monkey primary afferents, Journal of Neurophysiology 46 (1981 Dec) 1192–1203, https://doi.org/10.1152/jn.1981.46.6.1192.
[18] G.J. Gerling, G.W. Thomas, Fingerprint lines may not directly affect SA-I mechanoreceptor response, Somatosensory & Motor Research 25 (2008 Jan) 61–76, https://doi.org/10.1080/08990220701838996.
[19] R. Ellis, M. Qin, Singular-value and finite-element analysis of tactile shape recognition, in: Proceedings of the 1994 IEEE International Conference on Robotics and Automation, ROBOT-94, IEEE Comput. Soc. Press, 1994, pp. 2529–2535, https://doi.org/10.1109/robot.1994.351131.
[20] H. Zhang, N. Chen, Control of contact via tactile sensing, IEEE Transactions on Robotics and Automation 16 (2000) 482–495, https://doi.org/10.1109/70.880799.
[21] A. Caiti, G. Canepa, D. De Rossi, F. Germagnoli, G. Magenes, T. Parisini, Towards the realization of an artificial tactile system: fine-form discrimination by a tensorial tactile sensor array and neural inversion algorithms, IEEE Transactions on Systems, Man and Cybernetics 25 (1995 Jun) 933–946, https://doi.org/10.1109/21.384256.
[22] K.L. Johnson, Contact Mechanics, Cambridge University Press, Cambridge, England, 1985.
[23] R. Fearing, T. Binford, Using a cylindrical tactile sensor for determining curvature, IEEE Transactions on Robotics and Automation 7 (1991) 806–817, https://doi.org/10.1109/70.105389.
[24] S.P. Karason, M.A. Srinivasan, A.M. Annaswamy, Encoding and decoding of static information in tactile sensing systems, The International Journal of Robotics Research 18 (1999 Feb) 131–151, https://doi.org/10.1177/027836499901800202.
[25] G. Holzapfel, Nonlinear Solid Mechanics, a Continuum Approach for Engineering, John Wiley & Sons, 2000, second print.
[26] R.S. Fearing, J.M. Hollerbach, Basic solid mechanics for tactile sensing, The International Journal of Robotics Research 4 (1985) 40–54, https://doi.org/10.1177/027836498500400304.
[27] R.D. Howe, Tactile sensing and control of robotic manipulation, Advanced Robotics 8 (1993 Jan) 245–261, https://doi.org/10.1163/156855394x00356.
[28] J. Jurczyk, K. Loparo, Mathematical transforms and correlation techniques for object recognition using tactile data, IEEE Transactions on Robotics and Automation 5 (1989 Jun) 359–362, https://doi.org/10.1109/70.34771.
[29] S. Ricker, R. Ellis, 2-D finite-element models of tactile sensors, in: Proceedings IEEE International Conference on Robotics and Automation, vol. 1, 1993, pp. 941–947.
[30] P.S. Khalsa, R.M. Friedman, M.A. Srinivasan, R.H. Lamotte, Encoding of shape and orientation of objects indented into the monkey fingerpad by populations of slowly and rapidly adapting mechanoreceptors, Journal of Neurophysiology 79 (1998 Jun) 3238–3251, https://doi.org/10.1152/jn.1998.79.6.3238.
[31] R.S. Johansson, I. Birznieks, First spikes in ensembles of human tactile afferents code complex spatial fingertip events, Nature Neuroscience 7 (2004 Jan) 170–177, https://doi.org/10.1038/nn1177.

Part III

Sensing

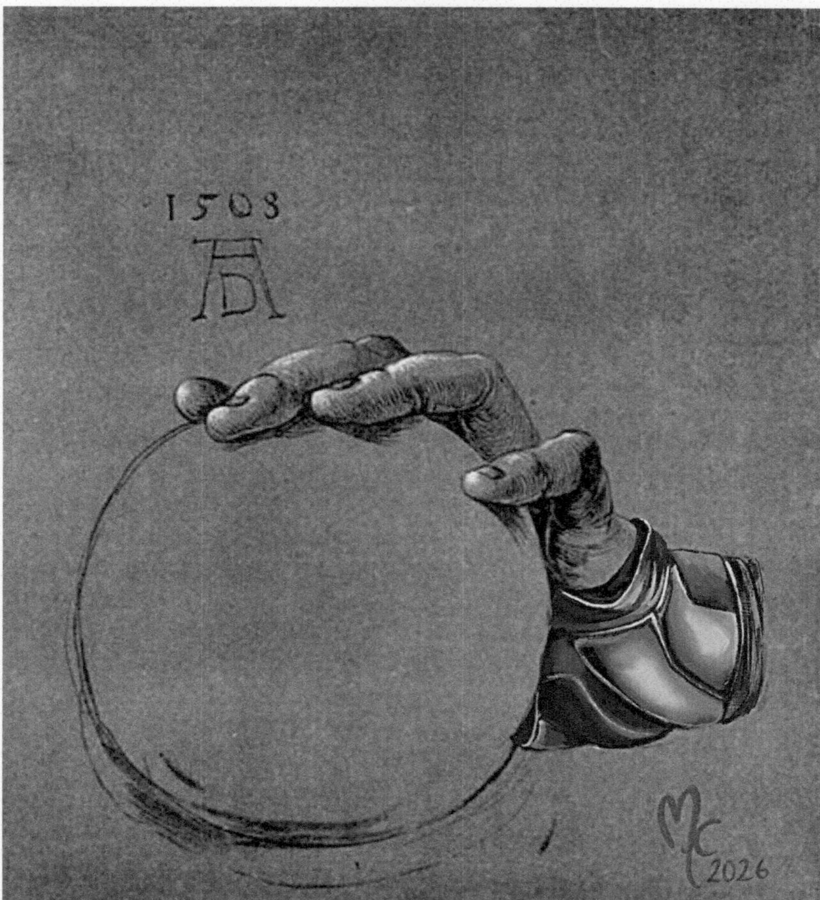

Chapter 7

Human and artificial tactile sensing

7.1 Introduction

An impaired or missing tactile sensing system will make dexterous grasping and manipulation difficult. In earlier studies when the fingers were anesthetized [1][2][3], the grasping and lifting of objects became difficult. These studies showed that the fingers applied less grip force on the object, and there were latencies in the grip forces that the fingers applied.

The successful grasping and manipulation of objects can be attributed to the human tactile sensing system. The forces and motions between an object and the fingers need to be controlled. The information on the contact interaction is provided by the tactile sensory system and is passed to the control system that triggers the sensorimotor programs of the human hand (see Fig. 1.2 and in [4][5]). Unlike the human eyes and the vision system, which can only indirectly infer an object's properties, the human tactile sensing system provides direct measurements of the object's physical properties, such as shape, surface details, hardness, temperature, vibration, mass, and friction [6].

The human sense of touch depends on four types of low-force threshold mechanoreceptors, which are either fast-adapting (FA) or slow-adapting (SA) [7][8]. These were further categorized into type I (small receptive fields) and type II (diffused receptive fields). FA I and SA I units are densely packed in the distal phalanges and give excellent spatial resolution and event detection. FA II and SA II units, on the other hand, are more responsive to the changing stimuli and long-lasting skin deformations on the skin surface [9][10].

For artificial tactile sensing, the technologies have advanced significantly. The typical transduction techniques that were reported in the literature involve resistive, capacitive, piezoelectric, and optical sensing technologies [11][12][13]. With known transduction techniques, sensors were designed to replicate how humans perceive touch, with the goal to give similar dexterity to robots and prosthetic hands [14][15].

The next section, Sec. 7.2, describes the biological foundations of touch by providing details on the four classes of human cutaneous mechanoreceptors. Sec. 7.3 then describes the typical transduction techniques in artificial tactile sensing. In Sec. 7.4, characteristics from both human and artificial tactile sensing are contrasted with respect to various classification schemes. Finally,

Humanizing Artificial Touch. https://doi.org/10.1016/B978-0-32-395519-5.00018-0

the last section summarizes the chapter and reinforces the benefits of biomimetic design philosophy to achieve optimized solutions for tactile sensing.

7.2 Human tactile sensing

Developed in the late 1960s [16], the microneurography technique made it possible to study the neurophysiological properties of the human mechanoreceptors. In this technique, a fine, insulated tungsten electrode is inserted percutaneously into the nerve and carefully manipulated until the uninsulated 5-20 μm tip picks up signals from a single afferent nerve fiber originating in an individual peripheral receptor. In the next subsections, some of the characteristics of the mechanoreceptors are described.

7.2.1 Types of human mechanoreceptors

Among all types of afferents supplying the human glabrous skin, the low-threshold, fast-conducting afferents are relevant for this book. They belong to one of two main groups [7][8]: the Slowly Adapting (SA) and Fast Adapting (FA) afferents. The SA afferents show a sustained response to indenting stimuli while the FA afferents respond to dynamic events. Both SA and FA can be further subdivided into Type I and Type II afferents on the basis of their receptive field (RF) properties. By definition, the RF of an afferent corresponds to the skin area in which indenting stimuli evoke action potentials. The detailed topography of RFs could be very complex [18]. Type I afferents possess small and well-delineated RFs whereas the RFs of Type II afferents are large with poorly defined borders [8].

The mechanoreceptors present in the glabrous skin of the human hand and their distinctive characteristics are summarized in Fig. 7.1. The behavior and mechanical characteristics of the mechanoreceptors offer a wealth of ideas for functional mimicry in the design of artificial tactile sensors. The reported values may also serve as useful benchmarks for characterization or for comparative evaluation of tactile sensors.

The columns with pink (light gray in print version) and green (pale gray in print version) backgrounds illustrate the adaptability, directional preference, and mechanical properties of mechanoreceptors. The mechanoreceptor adaptation illustrates the spike responses in response to perpendicular ramp-hold-return indentations of the skin for each unit type [9]. The fingertip shown with arrows pointing in different directions shows the directional preference of each mechanoreceptor in response to tangential forces applied to it [19]. The arrows originate from the location of stimulation and the white arrows show the resultant direction. Additional information on the activation threshold, frequency range, and their sensitivities are shown.

The leftmost column shows the receptive field, whereas the rightmost column illustrates the innervation density for Type I and Type II mechanoreceptor units. The Type I units possess limited and distinctly defined cutaneous receptive

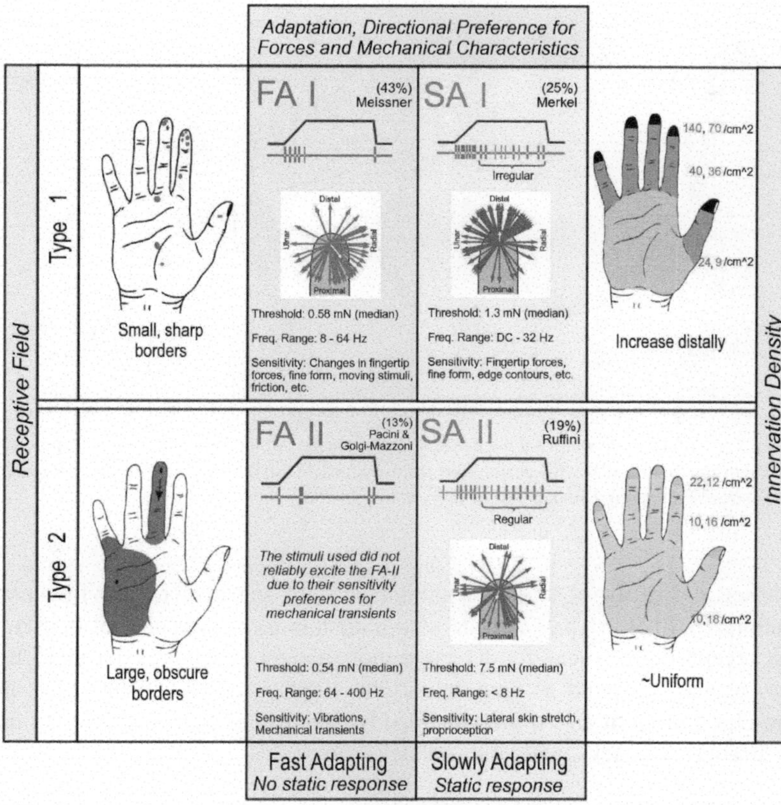

FIGURE 7.1 The types of mechanoreceptors in the glabrous skin of the human hand and some of their distinguishing properties. The vertical panels with pink (light gray in print version) and green (pale gray in print version) backgrounds present the mechanoreceptors' adaptation, directional preference and mechanical characteristics in a descending manner. Adapted from Johansson and Flanagan [17], with permission.

fields. The Type II units, characterized by lower and more uniform densities, exhibit responsiveness to more distant stimuli, possessing bigger and less precisely defined receptive fields.

The figures displayed on the rightmost column indicate the approximate density per cm^2 of the FA units (in red – medium gray in print version) and SA units (in green – dark gray in print version). For more detailed descriptions on the roles and functions of mechanoreceptors, the interested reader is referred to the source papers in Johansson and Westling [20], Johnson [21], Johansson and Flanagan [17]. Short descriptions are provided next.

7.2.1.1 Fast adapting, type 1 (FA I): Meissner corpuscles

FA afferents innervate Meissner corpuscles located just beneath the epidermis in dermal papillae of glabrous skin (Fig. 7.2). They respond transiently at

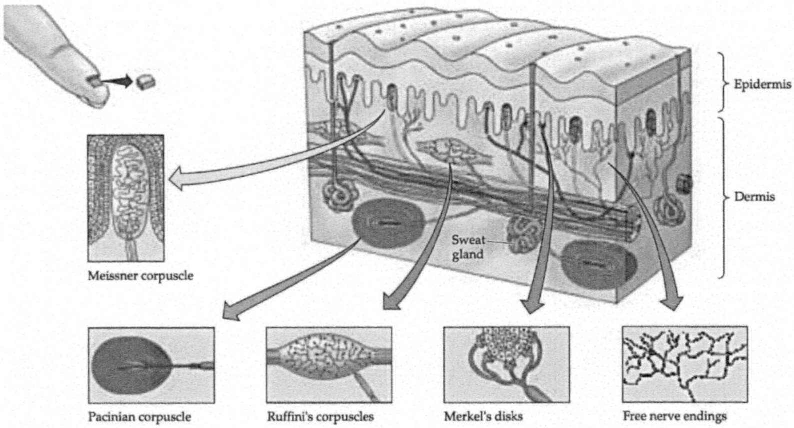

FIGURE 7.2 Mechanoreceptors at the hairless (glabrous) skin of the human fingertip. Reproduced from Augustine et al. [22], with permission from Oxford Publishing Limited through PLSclear. ©2004 Oxford Publishing Limited. All rights reserved.

the onset and offset of skin indentation and are sensitive to low-frequency vibrations (~40 Hz). They activate with a median force threshold of 0.58 mN. These receptors are crucial for detecting motion and fine textures during sliding touch, as they respond to skin vibrations elicited by movement. Meissner corpuscles consist of an axon surrounded by specialized Schwann-like laminar cells arranged in stacks, with a fibroblast capsule.

7.2.1.2 Slow adapting, type 1 (SA I): Merkel cells

SA I afferents densely innervate the skin, especially the fingertips (~100 per cm^2), and respond with sustained firing to static indentation proportional to indentation depth. They are highly sensitive to points, edges, and curvature due to selective sensitivity to strain energy density. Their receptive fields are about 2-3 mm in diameter, with spatial resolution down to 0.5 mm, enabling acute spatial neural images critical for form and texture perception. They have a median force threshold of 1.3 mN. Merkel cells, specialized epidermal cells, envelop the afferent endings, contributing to this selectivity.

7.2.1.3 Fast adapting, type 2 (FA II): Pacinian corpuscles (PC)

Pacinian corpuscles are large, onion-like structures located deep in the dermis and hypodermis. They are highly sensitive to high-frequency vibrations (20-1500 Hz, with peak of 200-400 Hz) and respond to pressure and vibratory stimuli with exquisite sensitivity to nanometer-scale deformations. They have a median force threshold of 0.54 mN. The corpuscle's layered lamellar cells and capsule act as mechanical filters protecting the nerve ending. PC afferents play a key role in detecting vibrations transmitted through objects and skin.

7.2.1.4 *Slow adapting, type 2 (SA II): Ruffini corpuscles*

SA II afferents innervate Ruffini corpuscles, spindle-shaped structures embedded in the dermis and connected to the collagen matrix. Their frequency range is less than 8 Hz and their median force threshold is 7.5 mN. They are selectively sensitive to skin stretch and contribute to the perception of skin deformation and force. These mechanoreceptors are sensitive for lateral skin stretch and proprioception.

7.2.2 Bio-inspired tactile sensing requirements

Listed below are the basic requirements of a bio-inspired tactile system that can fulfill the functional requirements of its biological counterpart with respect to a prototypical grasp and lift task. Longer reviews can be found in [23][24][25][26][27].

The artificial tactile sensory system of an anthropomorphic hand should provide basic information about:

1. contact and release between the fingertip and the object (Chap. 8);
2. contact between the object and the environment (Chap. 8);
3. slippage between the fingertip and the object (Chap. 9);
4. the local surface primitives at the contact point (Chap. 10), and
5. the overall object shape primitives (Chap. 11).

The first three requirements are necessary for simple precision grasping tasks. The last two requirements are necessary for tactile perception. For the details on the last two requirements, the interested reader is referred to Jenmalm et al. [28] for local shape and to Lederman and Klatzky [29] regarding the overall object shape. The next chapters in Part III – Sensing propose some of the potential solutions to address those requirements.

7.3 Artificial tactile sensing

Early definitions by Harmon [30] refer to tactile sensing as having skin-like properties with which areas of force-sensitive surfaces are capable of reporting graded signals and parallel patterns of touching. The definition has been broadly expanded to accept any property that is measured through contact like shape of an object, texture, temperature, hardness, moisture content as tactile sensing [12]. Formally, tactile sensing is defined as the process of determining physical properties and events through contact with objects in the world. A tactile sensor is a device or system that can measure a given property of an object or contact event through physical contact between the sensor and the object.

A distinction exists between intrinsic and extrinsic tactile sensing. Extrinsic sensors are devices that are distributed over the contact surfaces whereas intrinsic sensors, which are often located within the fingertip structure, refer to the derivation of contact data from force/torque sensors [11]. The idea of tactile sensing based on pure force/torque measurements was first proposed in

[31] and advanced by [32][33]. Extrinsic sensors were further classified as being in an array configuration or as single-valued components [12]. The majority of tactile sensing devices in the literature are based on extrinsic sensing.

In the study of the biological touch, the term tactile sensing was differentiated from the terms kinesthetic sensing and haptics [6]. While tactile refers to the processing of solely cutaneous information, the term kinesthetic refers to any input pertaining to the mechanical configuration of a joint, finger, hand or arm. The term haptics describes the processing of both cutaneous and kinesthetic inputs for perceiving and guiding actions within the real world. It is the means through which characteristics such as object shape, surface texture, stiffness and temperature are known. This terminology developed in the field of biological touch has also found its way to the field of robotics.

7.3.1 Types of tactile sensors

Tactile sensors for prosthetics and robot hands fall into several transduction families that convert mechanical contact into electrical signals. A schematic representation of tactile sensors and their possible embedding locations are shown in Fig. 7.3. Some of the typical tactile sensors are described below according to their transduction principle.

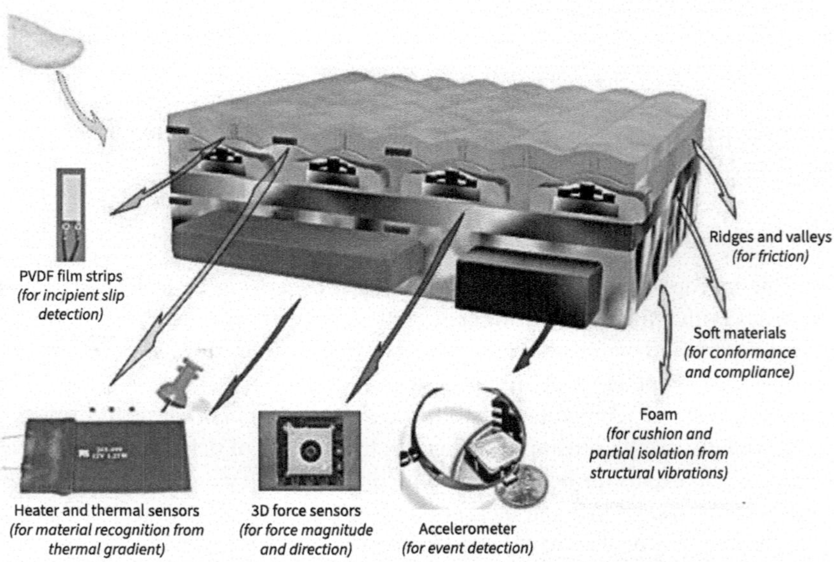

FIGURE 7.3 Schematic representation of tactile sensors embedded on artificial skin. Representative sensors that could address the requirements for tactile sensing for PVDF film strips for incipient slippage, heater and thermal sensors for thermal gradient, 3D force sensors for force magnitude and direction, and accelerometer for event detection.

7.3.1.1 Piezoresistive sensors

Piezoresistive sensors rely on resistance changes in a material when deformed; in a common form, the piezoresistive resistance is:

$$R = \frac{\rho L}{A} \tag{7.1}$$

where ρ is the material's resistivity, L is the length and A the cross-sectional area. The intrinsic piezoresistive gauge factor, GF, can be described as:

$$GF = \frac{\Delta\rho/\rho}{\epsilon_a} \tag{7.2}$$

where $\Delta\rho$ is the resistivity change and ϵ_a is the axial strain. This equation underpins Force Sensing Resistors (FSR) and conductive rubbers used in many robotic fingertips [34][35]. These sensors have advantages of being flexible and are easier to manufacture. They have the drawbacks of high hysteresis, nonlinearities in the results, and temperature sensitivity [13].

7.3.1.2 Capacitive sensors

Capacitive sensors measure contact by sensing the variations in capacitance. Capacitance is calculated as:

$$C = \varepsilon_r \varepsilon_0 \frac{A}{d} \tag{7.3}$$

where ε_r is the relative permittivity of the dielectric, ε_0 the vacuum permittivity, A the overlapping area and d the plate separation. In array configurations, mutual capacitance, C_{mutual}, between orthogonal electrodes is read to infer locations where touch occurred. The advantages of capacitive sensing include high bandwidth and often good linearity. Its disadvantages include the susceptibility to electromagnetic interference, temperature drift, and hysteresis [36][37].

7.3.1.3 Piezoelectric sensors

Piezoelectric sensors utilize the direct coupling between mechanical stress and electrical charge via the piezoelectric effect. The output voltage can be obtained as:

$$V = \frac{Q}{C} \tag{7.4}$$

where the output voltage relates to charge Q and capacitance C, with Q proportional to the applied stress. Polyvinylidene fluoride (PVDF) tactile sensors illustrate high bandwidth and dynamic range, though they are typically more effective for dynamic rather than static forces [38].

7.3.1.4 Optical sensors

Optical tactile sensors use light to map contact. Here, the deformations alter the light propagation or reflection, and cameras or photodiode arrays capture the resulting patterns. A GelSight-style sensor, for example, uses a transparent elastomer and a camera to reconstruct high-resolution surface textures [39]. An optical readout can be modeled as:

$$I = I_0 f(\delta) \tag{7.5}$$

where I is the detected light intensity, I_0 is the incident light, and $f(\delta)$ is a function dependent on the distance between the sensor and the object.

The tactile sensors of these types are often utilized on artificial fingers for grasping and manipulation tasks. They are used to detect normal and tangential forces, texture, temperature, and other object properties. The large-area sensors, otherwise known as electronic skins (e-skins), are another type of tactile sensors that are thin, flexible, stretchable, and can even have self-healing properties. This type is beyond the scope of this chapter.

7.3.2 Tactile sensor requirements

Harmon [30] proposed a set of sensor requirements in the early 1980s based on a questionnaire sent to manufacturers and robotic researchers. Some of these have remained as a rule-of-thumb specification to some researchers to date. These requirements are summarized as follows:

1. The sensor surface should be both compliant and durable;
2. Spatial resolution should be 1-2 mm;
3. A range of 50-200 tactile elements is acceptable;
4. The force threshold is 0.049 N, ideally 0.0098 N;
5. A dynamic range of 1000:1 is satisfactory;
6. The sensor must be stable, repeatable and without hysteresis;
7. The response must be monotonic, though not necessarily linear;
8. The time resolution for the sensor should be at least 100 Hz.

Jacobsen et al. [40] first pointed out that despite the progress in sensor fabrication in those days, no rigorous tactile sensing theory existed for specifying important system parameters such as sensor density, resolution, location and bandwidth. Several years later, Howe [14] again remarked that the field of tactile sensing lacked a comprehensive theory that defines the sensing requirements for various manipulation tasks. Due to this gap in tactile sensing theories, the designers commonly model systems after the human hand and the mechanoreceptors of the glabrous (hairless) skin.

It was noted, however, that the neurophysiological, psychophysical, and psychological knowledge about primate and human tactile sensing mechanisms was still incomplete [41][11]. Moreover, biological mechanoreceptors are both non-linear (including hysteretic) and time-varying, in contrast to the properties desired for practical engineering applications.

7.4 Comparisons: human vs artificial tactile sensors

7.4.1 Classification

Human FA and SA afferents adapt fast and slowly, respectively, to maintained indentation of the skin. The FAs respond to skin deformation changes whereas the SAs also show an ongoing response during periods of static skin deformation. The type I afferents (FA I and SA I) have small and well-delineated receptive fields. The receptive fields of the type II are large and poorly defined.

Comparisons between engineered sensors and the human mechanoreceptors can easily be found in the robotics literature. In [42], they suggested that polyvinylidene fluoride (PVDF) strips can be useful detectors of incipient slippage like the FA I of the human hand. Howe and Cutkosky [43] showed the event detection capability of accelerometers similar to FA II. As will be shown in the next chapter, Chap. 8, on-off contact sensors with their constant static response throughout the duration of fingertip-object contact are analogous to SA I mechanoreceptors.

Further simplifications can be made for artificial tactile systems by using the classification scheme of global and local sensors. The global sensors are sensitive to mechanical transients like vibrations. Likewise, the local sensors may also be sensitive to mechanical transients but must provide spatial information by means of its location and limited receptive field.

7.4.2 Sensor distribution

There are about 17,000 mechanoreceptive units innervating the glabrous part of the human hand [18][7]. The majority of these are due to the units with small and well-defined receptive fields, the FA I (43%) and SA I (25%). The FA II (13%) and SA II (19%) are almost evenly distributed over the whole glabrous skin area. The overall density was found to increase in the proximo-distal direction, i.e., starting from the base of the finger to the fingertip. Findings indicate that the density of the mechanoreceptors increases slightly from the palm to the proximal/middle parts of the fingers and then show an abrupt increase from the proximal/middle parts of the finger to the fingertip [7]. The relative densities in these three regions were 1 for the palm, 1.6 for the proximal and middle phalange, 4.2 for the distal phalange. The huge concentration of tactile units in the fingertips accounts for our ability to localize stimuli correctly on the skin surface and appreciate delicate details of objects that we touch and manipulate [9][8].

The human tactile system of the glabrous skin shows clear evidence of redundancy. In humans, a study by Gardner [44] shows that the number of myelinated fibers present in the second and the third decade of life is reduced by 5% per decade. This implies that the number of sensing units innervating the glabrous skin of a human hand is reduced from 17,000 to 13,000 in the eighth decade of life [7]. Although our tactile acuity decreases with time, our touch sensing capability never stops functioning. A redundant design thus provides a good safety net against common problems of engineered sensory systems (e.g., wiring disconnection, breakage).

For a tactile system mounted on a robotic hand, the Gifu Hand III hand [45] had 859 tactile contact points. It has a homogeneous distribution of sensitive areas along the palm and all of the fingers. The F-TAC Hand [46] demonstrated a 0.1 mm tactile spatial resolution and coverage of over 70% of the hand surface, with 10,000 taxels/cm^2 on the palm.

The human system can also provide optimization ideas for the distribution of tactile sensors. A non-homogeneous distribution of sensors can be implemented that has a high concentration on the fingertip compared to the other parts. However, this type of configuration may present a technical constraint for large and commercially available sensors. Since most artificial hand designers usually follow the anthropometry of the human hand, the distal phalanges of the robotic hand have the least dimensions compared to the other phalanges. The assembly of the tactile sensors in such a small space is a challenging task. Advances in microsensors and nanotechnology are ideal for this kind of task. Another variation is to implement a retina-like foveated design with high density of tactile elements in the region expected to have the most contact interactions, similar to [47][48].

7.4.3 Activation threshold

The activation threshold is the minimal force at which the mechanoreceptor responds to an indentation stimulus. The thresholds recorded for the tactile afferents are as follows: FA I (0.58 mN), SA I (1.3 mN), FA II (0.54 mN) and SA II (7.5 mN) [9]. Roughly 90% of the FA I and SA I units are excited by a 5 mN force and indentations of about 1 mm would excite practically all the FA I and SA I units [9]. Low activation threshold is important in the early determination of contact between the fingertips and the object. If the threshold is too high, contact may have long been initiated.

7.4.4 Receptive field and spatial resolution

Receptive field (RF) is a measure of the sensitive areas of a mechanoreceptor unit when a stimulus is delivered on the skin, often defined as the area within which an action potential can be evoked with a stimulus four times the activation threshold. The shapes of the RFs of human skin mechanoreceptors defined in this manner are roughly round or slightly oval. The reported values of the receptive fields were as follows: FA I (9.4 mm^2), SA I (11 mm^2), FA II (101.3 mm^2) and SA II (58.7 mm^2) [9]. These correspond to circular areas with diameters of 3.4 mm, 3.7 mm, 11.4 mm and 8.7 mm, respectively. Except for the FA I, there were no significant statistical differences between the sizes of the fields for the regions of the distal phalange, the rest of the finger and the palm. The receptive field of the FA I for the main parts of the finger is 12.6 mm^2 and 28.1 mm^2 for the palm.

Spatial resolution, on the other hand, is a measure of the finest spatial stimulus that could be discriminated reliably [49]. In humans, spatial discrimination depends on the RF sizes and sensor densities. Several methods have been proposed to measure the threshold of spatial resolution in humans. Among these, the two-point threshold is the most well-known. However, some problems are associated with its use. Some of these problems include setting an appropriate criterion for responding “two”, the extreme variability both within and between subjects and the ability of subjects to discriminate two points from one separation well below the two-point threshold [50].

Two alternative measures have been devised to overcome the limitations of the two-point test. The first is grating orientation. To obtain this measure, participants were presented with a grating consisting of alternating grooves and ridges. The grating was presented in one of two orientations that are at right angles to one another. The participant was asked to report the orientation. The gap size at the threshold was 0.84 mm [51]. The second measure is gap detection. In this measure, a surface with or without gap is pressed towards the skin. The gap varies in size and the task of the human subject is to indicate whether there is a gap or not. The gap size at the threshold of the human perception was 0.87 mm [51].

Receptive field and spatial resolution measurements over a viscoelastic skin covering are rarely reported. The spatial resolution is often reported as the measurable distance between two sensing elements without the skin covering, i.e., element pitch. The suggested spatial resolution in robotic tactile sensing is 1–2 mm [30][52]. Designs for tactile imagers, however, have reported high spatial resolutions of 300 μm for a traction stress sensor array [34] and 50 μm for a microelectromechanical system (MEMS) fingerprint sensor [53].

7.4.5 Frequency range

The frequency range is important for detecting dynamic events like slippage. In humans, the responses of mechanoreceptors to perpendicular sinusoidal skin displacements were recorded [54]. The FA I units responded most easily within the range of 8-64 Hz, the SA I units most easily between 2-32 Hz, the FA II units most easily between 64-400 Hz and the SA II units most easily below 8 Hz.

In robotics, there are at least three competing ideas on slippage sensing in the state of the art, intrinsic force sensors [32][33][26], accelerometers [14][55] and piezoelectric transducers [56][57]. To be applicable as a bio-inspired dynamic sensor, the bandwidth has to be sufficient to correspond to both SA and FA tactile afferents, i.e., DC – 400 Hz.

7.4.6 Wiring

Despite the significant advances in sensor miniaturization and performance, the electrical wiring remains a weakness of robotic tactile systems. Wires impede mobility, are obtrusive and are vulnerable to breakage [58]. On the contrary, human biology has implemented a superior solution. The nervous system is composed of a complex network of "wires". Each mechanoreceptor is connected to a nerve fiber that has a size of 6-14 μm and there are 50,000 of these in the nerve.

Some solutions in tactile sensing include wireless tactile sensors [59][60] and optical signal transmission [61]. Another approach is by integrating the processing electronics with the sensors [62]. A stretchable wire based on gold film was also described [63]. The film has a thickness of 25 nanometers and can stretch by at least 15% in the rubbery silicone membrane in which they are embedded.

7.4.7 Force preference

Precise control of fingertip forces and movements is critical for dexterous manipulation [10]. Information about the direction of the fingertip forces provides early information related to changes in the object's mechanical behavior and about contact conditions between an object and the hand. Likewise, in tasks in which humans restrain objects that are subjected to unpredictable external tangential load forces, the automatic grip reactions used to stabilize the object depend on the direction of load.

The sensitivity of the various types of tactile afferents to force directions was tested using standardized force stimuli delivered to the fingertip at magnitudes and rates compatible with those arising in everyday manipulative tasks [10]. The force delivered was 4 N while sites selected coincide with the contact region that humans prefer to use in goal-directed fine manipulation of small objects [64]. The directional preferences for force are shown on Fig. 7.1.

Birznieks et al. [10] hypothesized several conceivable mechanisms for extracting directional information from tactile afferents from the fingertips. One of these is that by monitoring the balance between the input across the population of SA I afferents and that across the SA II and FA I afferents, the brain could obtain cues concerning tangential force in distal and proximal directions.

7.5 Summary

Understanding the basic characteristics of the human mechanoreceptors is the first step to humanizing the artificial tactile sensing system of humanoid robots and prosthetic arms.

The complex sensory system of the human hand represents a supremely efficient model against which most prosthetic and humanoid hands are compared.

Tactile sensor designers have often referred to the human mechanoreceptors at the glabrous (hairless) part of the hand to draw inspiration [65][66][67]. The FA I (Meissner Corpuscles) and SA I (Merkel Discs) have smaller receptive areas. They are mostly concentrated near the skin surface and are responsible for detecting dynamic and static contact interactions. The FA II (Pacinian Corpuscles) and SA II (Ruffini Corpuscles) have larger receptive areas and are concentrated more deeply beneath the fingertip's skin surface. The FA II mechanoreceptors are sensitive to high frequency vibrations while the SA IIs are sensitive to skin stretch and contribute to the perception of skin deformation and forces.

These receptors possess the capability to perceive various physical characteristics and exhibit exceptional accuracy in responding to mechanical stimuli. With its ability to finely adjust grip forces and detect small changes in object attributes, the network of mechanoreceptors in the human tactile system can be advantageous for developing sophisticated artificial tactile sensing systems. The previous sections described some of the transduction principles for resistive, capacitive, piezoelectric, and optical sensing techniques to replicate some aspects of the mechanoreceptors. Drawing from the literature, some of the relevant characteristics of the human mechanoreceptors and artificial tactile sensors were compared. The insights on sensing characteristics on sensor distribution, activation thresholds, receptive fields, spatial resolution, frequency ranges, and force preferences could be useful for the designers of tactile sensing systems.

Tactile sensing is indispensable for dexterous grasping and manipulation. The fingertip-object interface is the best location to provide information regarding critical properties and events at the region of contact, such as forces and motions [14]. Moreover, by touch it is possible to directly measure physical properties of an object such as shape, surface details, hardness, temperature, vibration, moisture content, mass and friction [30][43][11]. By vision such properties can at best be deduced indirectly [11], and this is only possible when the target object is not occluded.

Knowing the functional aspects of biological sensors that meet the requirements of the basic grasping tasks provides meaningful directions in the design and development of artificial tactile systems that could replicate the nuanced and adaptive responses of the human tactile sensing system.

References

[1] R.S. Johansson, C. Häger, L. Bäckström, Somatosensory control of precision grip during unpredictable pulling loads: III. Impairments during digital anesthesia, Experimental Brain Research 89 (1992) 204–213, https://doi.org/10.1007/bf00229017.

[2] J. Monzée, Y. Lamarre, A.M. Smith, The effects of digital anesthesia on force control using a precision grip, Journal of Neurophysiology 89 (2003 Feb) 672–683, https://doi.org/10.1152/jn.00434.2001.

[3] A. Carteron, K. McPartlan, C. Gioeli, E. Reid, M. Turturro, B. Hahn, C. Benson, Zhang W. Temporary, Nerve block at selected digits revealed hand motor deficits in grasping tasks, Frontiers in Human Neuroscience 10 (2016 Nov), https://doi.org/10.3389/fnhum.2016.00596.

[4] R.S. Johansson, Sensory Input and Control of Grip, Wiley, 2007, pp. 45–63, https://doi.org/10.1002/9780470515563.ch4.

[5] C. Laschi, G. Asuni, E. Guglielmelli, G. Teti, R. Johansson, H. Konosu, Z. Wasik, M.C. Carrozza, P. Dario, A bio-inspired predictive sensory-motor coordination scheme for robot reaching and preshaping, Autonomous Robots 25 (2007) 85–101, https://doi.org/10.1007/s10514-007-9065-4.

[6] S.J. Lederman, D.T. Pawluk, Lessons from the Study of Biological Touch for Robotic Tactile Sensing, in: H.R. Nicholls (Ed.), Advanced Tactile Sensing for Robotics, in: World Scientific Series in Robotics and Automated Systems, vol. 5, World Scientific, Singapore, 1992, pp. 151–192, https://doi.org/10.1142/9789814355964_0008.

[7] R. Johansson, A. Vallbo, Tactile sensibility in the human hand: relative and absolute densities of four types of mechanoreceptive units in glabrous skin, Journal of Physiology 286 (1979) 283–300, https://doi.org/10.1113/jphysiol.1979.sp012619.

[8] Å.B. Vallbo, R.S. Johansson, Properties of cutaneous mechanoreceptors in the human hand related to touch sensation, Human Neurobiology 3 (1) (1984) 3–14.

[9] R.S. Johansson, A.B. Vallbo, Spatial properties of the population of mechanoreceptive units in the glabrous skin of the human hand, Brain Research 184 (1980) 353–366, https://doi.org/10.1016/0006-8993(80)90804-5.

[10] I. Birznieks, P. Jenmalm, A.W. Goodwin, R.S. Johansson, Encoding of direction of fingertip forces by human tactile afferents, The Journal of Neuroscience 21 (2001) 8222–8237, https://doi.org/10.1523/jneurosci.21-20-08222.2001.

[11] Dario P. Tactile, Sensing: technology and applications, Sensors and Actuators. A, Physical 251 (6) (1991) 25–27, https://doi.org/10.1016/0924-4247(91)87001-J.

[12] M. Lee, Nicholls H. Review, Article tactile sensing for mechatronics—a state of the art survey, Mechatronics 9 (1999 Feb) 1–31, https://doi.org/10.1016/s0957-4158(98)00045-2.

[13] R. Dahiya, G. Metta, M. Valle, Sandini G. Tactile, Sensing—from humans to humanoids, IEEE Transactions on Robotics 26 (2010 Feb) 1–20, https://doi.org/10.1109/tro.2009.2033627.

[14] R.D. Howe, Tactile sensing and control of robotic manipulation, Advanced Robotics 8 (1993 Jan) 245–261, https://doi.org/10.1163/156855394x00356.

[15] M.C. Carrozza, P. Dario, L.P.S. Jay, Micromechatronics in surgery, Transactions of the Institute of Measurement and Control 25 (2003) 309–327, https://doi.org/10.1191/0142331203tm089oa.

[16] Å.B. Vallbo, K.E. Hagbarth, Activity from skin mechanoreceptors recorded percutaneously in awake human subjects, Experimental Neurology 21 (1968) 270–289, https://doi.org/10.1016/0014-4886(68)90041-1.

[17] R. Johansson, J. Flanagan, 6.05 – tactile sensory control of object manipulation in humans, in: R.H. Masland, T.D. Albright, T.D. Albright, R.H. Masland, P. Dallos, D. Oertel, S. Firestein, G.K. Beauchamp, M. Catherine Bushnell, A.I. Basbaum, J.H. Kaas, E.P. Gardner (Eds.), The Senses: A Comprehensive Reference, Academic Press, New York, 2008, pp. 67–86.

[18] R.S. Johansson, Tactile sensibility in the human hand: receptive field characteristics of mechanoreceptive units in the glabrous skin area, Journal of Physiology 281 (1978) 101–125, https://doi.org/10.1113/jphysiol.1978.sp012411.

[19] R.S. Johansson, I. Birznieks, First spikes in ensembles of human tactile afferents code complex spatial fingertip events, Nature Neuroscience 7 (2004 Jan) 170–177, https://doi.org/10.1038/nn1177.

[20] R.S. Johansson, G. Westling, Roles of glabrous skin receptors and sensorimotor memory in automatic control of precision grip when lifting rougher or more slippery objects, Experimental Brain Research 56 (1984) 550–564, https://doi.org/10.1007/bf00237997.

[21] K.O. Johnson, The roles and functions of cutaneous mechanoreceptors, Current Opinion in Neurobiology 11 (2001) 455–461, https://doi.org/10.1016/s0959-4388(00)00234-8.

[22] G.J. Augustine, D. Fitzpatrick, Purves D. Neuroscience, 3rd ed., Sinauer Associates, Sunderland, MA, 2004.

[23] M. Carrozza, P. Dario, F. Vecchi, S. Roccella, M. Zecca, F. Sebastiani, The Cyberhand: On the design of a cybernetic prosthetic hand intended to be interfaced to the peripheral nervous system, in: Proceedings 2003 IEEE/RSJ International Conference on Intelligent Robots and Systems (IROS 2003), vol. 3, IROS-03, IEEE, 2003, pp. 2642–2647, https://doi.org/10.1109/iros.2003.1249269.

[24] M. Carrozza, C. Suppo, F. Sebastiani, B. Massa, F. Vecchi, R. Lazzarini, M. Cutkosky, P. Dario, The SPRING hand: development of a self-adaptive prosthesis for restoring natural grasping, Autonomous Robots 16 (2004) 125–141, https://doi.org/10.1023/b:auro.0000016863.48502.98.

[25] M.C. Carrozza, G. Cappiello, S. Micera, B.B. Edin, L. Beccai, C. Cipriani, Design of a cybernetic hand for perception and action, Biological Cybernetics 95 (2006), https://doi.org/10.1007/s00422-006-0124-2.

[26] B. Edin, L. Beccai, L. Ascari, S. Roccella, J. Cabibihan, M. Carrozza, Bio-inspired approach for the design and characterization of a tactile sensory system for a cybernetic prosthetic hand, in: Proceedings 2006 IEEE International Conference on Robotics and Automation, ICRA 2006, 2006, pp. 1354–1358, https://doi.org/10.1109/robot.2006.1641897.

[27] B. Edin, L. Ascari, L. Beccai, S. Roccella, J.J. Cabibihan, M. Carrozza, Bio-inspired sensorization of a biomechatronic robot hand for the grasp-and-lift task, Brain Research Bulletin 75 (2008) 785–795, https://doi.org/10.1016/j.brainresbull.2008.01.017.

[28] P. Jenmalm, I. Birznieks, A.W. Goodwin, R.S. Johansson, Influence of object shape on responses of human tactile afferents under conditions characteristic of manipulation, European Journal of Neuroscience 18 (2003) 164–176, https://doi.org/10.1046/j.1460-9568.2003.02721.x.

[29] S.J. Lederman, R.L. Klatzky, Hand movements: a window into haptic object recognition, Cognitive Psychology 19 (1987) 342–368, https://doi.org/10.1016/0010-0285(87)90008-9.

[30] L.D. Harmon, Automated tactile sensing, The International Journal of Robotics Research 1 (1982 Jun) 3–32, https://doi.org/10.1177/027836498200100201.

[31] J.K. Salisbury, Interpretation of Contact Geometries from Force Measurements, Proceedings of the IEEE International Conference on Robotics and Automation (ICRA), IEEE, Atlanta, GA, USA, 1984, https://doi.org/10.1109/robot.1984.1087180.

[32] A. Bicchi, J.K. Salisbury, P. Dario, Augmentation of grasp robustness using intrinsic tactile sensing, in: Proceedings of the 1989 IEEE International Conference on Robotics and Automation (ICRA), 1989, pp. 302–307, https://doi.org/10.1109/ROBOT.1989.100005.

[33] A. Bicchi, Intrinsic contact sensing for soft fingers, in: Proceedings of the 1990 IEEE International Conference on Robotics and Automation (ICRA), Cincinnati, OH, 1990, pp. 968–973, https://doi.org/10.1109/ROBOT.1990.126117.

[34] B.J. Kane, M.R. Cutkosky, G.T.A. Kovacs, A traction stress sensor array for use in high-resolution robotic tactile imaging, Journal of Microelectromechanical Systems 9 (2000) 425–434, https://doi.org/10.1109/84.896763.

[35] M. Shimojo, A. Namiki, M. Ishikawa, R. Makino, K. Mabuchi, A tactile sensor sheet using pressure conductive rubber with electrical-wires stitched method, IEEE Sensors Journal 4 (2004) 589–596, https://doi.org/10.1109/JSEN.2004.833152.

[36] H.K. Lee, S.I. Chang, Yoon E. Dual-Mode, Capacitive proximity sensor for robot application: implementation of tactile and proximity sensing capability on a single polymer platform using shared electrodes, IEEE Sensors Journal 9 (2009) 1748–1755, https://doi.org/10.1109/JSEN.2009.2030660.

[37] R.S. Dahiya, P. Mittendorfer, M. Valle, G. Cheng, V.J. Lumelsky, Directions toward effective utilization of tactile skin: a review, IEEE Sensors Journal 13 (2013) 4121–4138, https://doi.org/10.1109/JSEN.2013.2279056.

[38] D. Göger, N. Ecker, Wörn H. Tactile, Sensor and Algorithm to Detect Slip in Robot Grasping Processes, Proceedings of the IEEE International Conference on Robotics and Biomimetics (ROBIO 2008), IEEE, 2009, pp. 1480–1485.

[39] R. Li, E.H. Adelson, Sensing and recognizing surface textures using a GelSight sensor, in: Proceedings of the IEEE Conference on Computer Vision and Pattern Recognition (CVPR), IEEE, 2013, pp. 1241–1247, https://doi.org/10.1109/CVPR.2013.164.
[40] S. Jacobsen, I. McCammon, K. Biggers, R. Phillips, Tactile Sensing System Design Issues in Machine Manipulation, Proceedings. 1987 IEEE International Conference on Robotics and Automation, vol. 4, Institute of Electrical and Electronics Engineers, 1987, pp. 2087–2096, https://doi.org/10.1109/robot.1987.1087755.
[41] P. Dario, D. De Rossi, Tactile Sensors and the Gripping Challenge, IEEE Spectrum, 1985, pp. 46–52, https://doi.org/10.1109/MSPEC.1985.6370785.
[42] I. Fujimoto, Y. Yamada, T. Maeno, T. Morizono, Y. Umetani, Development of artificial finger skin to detect incipient slip for realization of static friction sensation, in: Proceedings of the IEEE International Conference on Multisensor Fusion and Integration for Intelligent Systems (MFI), 2003, pp. 15–21, https://doi.org/10.1109/MFI-2003.2003.1232571.
[43] R.D. Howe, Cutkosky MR. Sensing, Skin acceleration for slip and texture perception, in: Proceedings of the IEEE International Conference on Robotics and Automation (ICRA), IEEE, 1989, pp. 145–150, https://doi.org/10.1109/ROBOT.1989.99981.
[44] E. Gardner, Decrease in human neurones with age, The Anatomical Record 77 (1940) 529–536, https://doi.org/10.1002/ar.1090770409.
[45] T. Mouri, H. Kawasaki, K. Yoshikawa, J. Takai, S. Ito, Anthropomorphic Robot Hand: Gifu hand III, in: Proceedings of the International Conference on Control, Automation and Systems (ICCAS'02), Jeonbuk, Korea, 2002, pp. 1288–1293.
[46] Z. Zhao, W. Li, Y. Li, T. Liu, B. Li, M. Wang, K. Du, H. Liu, Y. Zhu, Q. Wang, K. Althoefer, S.C. Zhu, Embedding high-resolution touch across robotic hands enables adaptive human-like grasping, Nature Machine Intelligence 7 (2025 Jun) 889–900, https://doi.org/10.1038/s42256-025-01053-3.
[47] P. Dario, R. Lazzarini, R. Magni, S.R. Oh, An integrated miniature fingertip sensor, in: Proceedings of the 7th International Symposium on Micro Machine and Human Science (MHS'96), Nagoya, Japan, 1996, pp. 91–97.
[48] J. Carpaneto, S. Micera, F. Zaccone, F. Vecchi, P. Dario, A sensorized Thumb for force closed-loop control of hand neuroprostheses, IEEE Transactions on Neural Systems and Rehabilitation Engineering 11 (2003) 346–353, https://doi.org/10.1109/TNSRE.2003.819938.
[49] R.W. Van Boven, K.O. Johnson, The limit of tactile spatial resolution in humans: grating orientation discrimination at the lip, tongue, and finger, Neurology 44 (1994) 2361, https://doi.org/10.1212/wnl.44.12.2361.
[50] J.C. Craig, K.O. Johnson, The two-point threshold: not a measure of tactile spatial resolution, Current Directions in Psychological Science 9 (2000) 29–32.
[51] K.O. Johnson, J.R. Phillips, Tactile spatial resolution. I. Two-point discrimination, gap detection, grating resolution, and letter recognition, Journal of Neurophysiology 46 (1981) 1177–1192, https://doi.org/10.1152/jn.1981.46.6.1177.
[52] M.T. Mason, J.K. Salisbury Jr., Robot Hands and the Mechanics of Manipulation, MIT Press, Cambridge, MA, 1985.
[53] N. Sato, K. Machida, H. Morimura, S. Shigematsu, K. Kudou, M. Yano, H. Kyuragi, MEMS Fingerprint Sensor immune to various finger surface conditions, IEEE Transactions on Electron Devices 50 (2003) 1109–1116, https://doi.org/10.1109/TED.2003.812490.
[54] R.S. Johansson, U. Landström, R. Lundström, Responses of mechanoreceptive afferent units in the glabrous skin of the human hand to sinusoidal skin displacements, Brain Research 244 (1982) 17–25, https://doi.org/10.1016/0006-8993(82)90899-X.
[55] M. Tremblay, M. Cutkosky, Estimating friction using incipient slip sensing during a manipulation task, in: Proceedings IEEE International Conference on Robotics and Automation, IEEE Comput. Soc. Press, 1993, pp. 429–434.
[56] J. Son, E. Monteverde, R. Howe, A tactile sensor for localizing transient events in manipulation, in: Proceedings of the 1994 IEEE International Conference on Robotics and Automation. ROBOT-94, IEEE Comput. Soc. Press, 1994, pp. 471–476, https://doi.org/10.1109/robot.1994.351253.

[57] J. Sone, Fingertip tactile sensation via piezoelectric micromachined ultrasonic transducers with an amplified interface, Scientific Reports 14 (2024 Feb), https://doi.org/10.1038/s41598-024-52630-2.

[58] I.D. McCammon, S.C. Jacobsen, Tactile Sensing and Control for the Utah/MIT Hand. Dextrous Robot Hands, Springer, New York, 1990, pp. 239–266, https://doi.org/10.1007/978-1-4613-8974-3_11.

[59] H. Shinoda, H. Oasa, Wireless tactile sensing element using stress-sensitive resonator, IEEE/ASME Transactions on Mechatronics 5 (2000) 258–265, https://doi.org/10.1109/3516.868917.

[60] M. Hakozaki, H. Shinoda, Digital tactile sensing elements communicating through conductive skin layers, in: Proceedings 2002 IEEE International Conference on Robotics and Automation (Cat. No. 02CH37292). vol. 4. ROBOT-02, IEEE, 2002, pp. 3813–3817, https://doi.org/10.1109/robot.2002.1014314.

[61] K. Yamada, K. Goto, Y. Nakajima, N. Koshida, H. Shinoda, A Sensor Skin Using Wire-Free Tactile Sensing Elements Based on Optical Connection, in: Proceedings of the 41st SICE Annual Conference, (SICE 2002) Osaka, Japan, IEEE, 2002 Aug, pp. 131–134, https://doi.org/10.1109/SICE.2002.1195198.

[62] J. Butterfass, M. Grebenstein, H. Liu, G. Hirzinger, DLR-Hand II: Next Generation of a Dextrous Robot Hand, Proceedings 2001 ICRA. IEEE International Conference on Robotics and Automation, vol. 1, IEEE, 2001, pp. 109–114, https://doi.org/10.1109/robot.2001.932538.

[63] S.P. Lacour, S. Wagner, Z. Huang, Z. Suo, Stretchable gold conductors on elastomeric substrates, Applied Physics Letters 82 (2003) 2404–2406, https://doi.org/10.1063/1.1565683.

[64] M.I. Christel, S. Kitzel, Niemitz C. How, Precisely do bonobos (pan paniscus) grasp small objects?, Folia Primatologica 19 (1998) 165–194, https://doi.org/10.1023/A:1020319313219.

[65] D. Yamada, T. Maeno, Y. Yamada, Artificial finger skin having ridges and distributed tactile sensors used for grasp force control, in: Proceedings 2001 IEEE/RSJ International Conference on Intelligent Robots and Systems. IROS-01, IEEE, 2001, pp. 686–691, https://doi.org/10.1109/iros.2001.976249.

[66] P. Dario, C. Laschi, S. Micera, F. Vecchi, M. Zecca, A. Menciassi, B. Mazzolai, M.C. Carrozza, Biologically-Inspired Microfabricated Force and Position Mechano-Sensors. Sensors and Sensing in Biology and Engineering, Springer, Vienna, 2003, pp. 109–125, https://doi.org/10.1007/978-3-7091-6025-1_8.

[67] J. Dargahi, S. Najarian, Human tactile perception as a standard for artificial tactile sensing—a review, The International Journal of Medical Robotics and Computer Assisted Surgery 1 (2004) 23–35, https://doi.org/10.1002/rcs.3.

Chapter 8

Sensing the fingertip, object, and environment contact

8.1 Introduction

Most of the artificial hands for humanoids and social robots have been compared with the human hand and its complex sensory system. Specifically, tactile sensor designers have often referred to the human mechanoreceptors at the glabrous part of the hand to draw inspiration [1][2][3][4]. However, to mimic the human hand's mechanoreceptors in detail may be ineffective, certainly costly and constrained by the state of technology.

The developments in the field of tactile sensors have been focused on the individual sensor components rather than as a complete system for tactile sensing [5]. A bioinspired tactile sensory system can be developed at two levels of abstraction: at a component level and at a system level. At the component level, the characteristics of the human mechanoreceptors were used as the guide. The tactile sensor components have been designed to have some functional resemblance with their biological counterparts in terms of activation threshold and frequency response.

At the system level, the sensor components can be designed to have a response comparable to human mechanoreceptors in the grasp-lift-replace task. The rationale is that most of the activities of our daily living can be characterized as having to reach an object, lift it and place it back to a target location. Extensive information is already available on how the human mechanoreceptors behave during this task [6][7], which could then be implemented in an artificial tactile sensing system.

The complexity of human grasping and manipulation has inspired extensive research in robotics, where the design of robotic hands increasingly incorporates multiple fingers, advanced sensing technologies, and sophisticated control algorithms to achieve dexterity comparable to that of humans [8][9].

This chapter extends the concepts discussed in the earlier chapter. The next section, Sec. 8.2, talks about the fingertip and object contact. Human grasping is discussed in light of the sequential and parallel coordination that it executes. Robot grasping is discussed in terms of the contact modeling at the robot fingertip and a surface. Sec. 8.3 extends the discussion to the interaction of the fingertip and the object, and the scenario when an object is lifted from a table. Likewise, the discussion is in consideration of both the human grasping and

Humanizing Artificial Touch. https://doi.org/10.1016/B978-0-32-395519-5.00019-2

contact modeling for a robot in an object lifting task. Sec. 8.4 presents a case study that ties both sections together. Lastly, Sec. 8.5 gives a summary for this chapter.

8.2 Human sensing for fingertip, object, and environment contact

The information from the sensory input of the glabrous part of the hand is crucial for precise grasping and manipulation. The most primitive task of the human tactile system is to provide contact and release information. Once adequate contact has been established, the motor commands are released for further manipulative actions. Furthermore, the tactile input from the skin provides information about object properties (i.e., friction, surface structures, shape, etc.) that are useful for adjusting the grip force. This ensures that an optimal force is applied, which is sufficient to prevent slippage but not so large as to crush the object.

Lifting an object and placing it back in the environment represents a basic human grasping and manipulation task [6]. We lift an object from a table, hold it in the air and replace it on the table.

8.2.1 Sequential coordination

The series of phases is shown in Fig. 8.1a. Distinct burst discharges in the tactile afferents correspond to the phase transitions. The responses for the following phases are obligatory [10]:

(a) initial responses appearing in the pre-load phase (phase a);
(b) FA II responses at the start of the lifting movement (phase b);
(c) transitional phase (phase c);
(d) static phase (phase d);
(e) replacement phase, burst responses shown again for the FA II upon the object's contact with the table (phase e);
(f) delay (phase f); and
(g) at the end of the unloading phase when the object is released (phase g).

The Central Nervous System (CNS) uses the information on the moment of contact or release to confirm that adequate contact is present before launching or terminating the motor commands [12]. The initial contact and release are sensed by SA I, FA I and FA II units. Spatially accurate information is required for determining which parts of the finger are engaged or disengaged. Moreover, the accurate spatial information is important during the manipulation of small objects for precise control of the directions of the various forces with reference to the contact areas, during tasks such as button insertion and tying shoelaces.

Distinct burst responses of FA II are triggered at the start and termination of the object's vertical movement. The vibration from the small lateral sliding movements between the object and the table can cause vigorous responses [12].

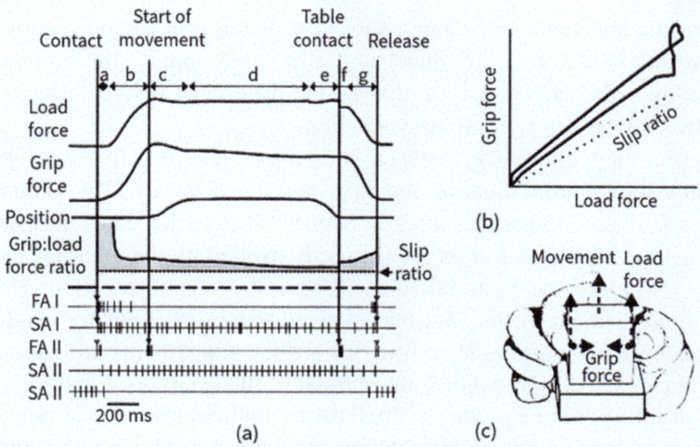

FIGURE 8.1 Schematic diagrams of the grasp-lift-replace task with the pinch grip. (a) Sequential coordination of the lifting task. The sensor data from an instrumented box showing the load force, grip force, and the position of the box. The task can be subdivided into various phases a-g, representing how the box is grasped, lifted, and replaced on the table. The lower panel shows the bursts of mechanoreceptor responses, with the FA II showing the instance when the object has been lifted from the table and returned to the table. (b) Parallel coordination in the lifting task. The grip force and load force showing a near-parallel increase during the task. The slip ratio is the limit for safe grasping to prevent an object from slipping from the grasp. (c) The experimental task showing the instrumented box during its vertical movement from the table. Reproduced from Edin et al. [11], with permission.

The motor drive that increases the grip and load forces is normally terminated following the FA II responses at the start of the movement. For a smooth vertical movement, the lifting drive must be decreased and appropriately adjusted to match the weight of the object [13]. At the termination, programmed motor commands were also released for the unloading phases. The need for sensory input to trigger these changes in the motor output is obvious when considering lifting conditions in which the weight of the object or the exact position of the support cannot be adequately predicted [10]. The CNS relies on these signals for quick, reliable and accurate triggering of the lifting events.

With anesthesia and without visual control, the phase from contact to the initiation of the load phase is significantly prolonged, presumably because contact has not been confirmed by the mechanoreceptors [14]. Likewise, under digital anesthesia, subjects may fail to notice that a lift-off takes place and accordingly continue to increase the vertical force until some other sensory system indicates the lift-off [15].

8.2.2 Parallel coordination

With the exception of the loading phase, it can be observed that the load and grip force increase and decrease in parallel (Figs. 8.1b and 8.1c). Westling and Johansson [10] showed that the grip:load force ratio has to be above a minimal level, i.e., the slip ratio, to ensure a safe grip. If the grip force is too low with respect to the load force, the object will slip. Accordingly, the human tactile system adjusts the safety margin, that is, the difference between the grip:load force ratio and the slip ratio, to prevent slippage.

The frictional condition between the object and the fingertip interface determines the coordination of the grip and load forces. The influence of material's surface commences already about 100 ms after the object has been initially gripped. The FA I units are primarily excited by the material's surface. The more slippery the material, the stronger the response is [6]. However, the initial adjustment to the friction is sometimes insufficient for establishing an adequate safety margin. Therefore, secondary adjustments are necessary to upgrade the grip:load force ratio later in the lift. The small slips are rarely felt by the human subject but responses in FA I, FA II and SA I units trigger an increase in the force ratio about 70 ms after the onset of slip thus preventing future slips. The process of frictional adaptation in the human tactile system represents an efficient way of producing flexibility with a minimum amount of reorganization [15]. Only one parameter needs to be regulated by the sensory information, the grip:load force ratio.

8.2.3 Force coordination

The process of lifting an object involves the coordination of two important forces: the grip force (GF) and the load force (LF) [6]. The GF needs to be kept sufficiently high to couple the LF with the object through friction between the fingers and the object to lift the object. An optimal LF is required to overcome the object's weight and lift it off the ground [14]. In humans, the GF and LF are thought to be generated in parallel by different subsystems in the brain [16][17].

The coordination between the GF and LF is crucial, as it determines the object's ability to be lifted and held in a stable manner [18][19]. In the lifting phase, there is a preload phase where the fingers first touch the object, followed by the load phase where the LF overcomes the object's weight and the object starts to move [20].

During the lifting process, the nervous system uses sensory feedback to rapidly and dynamically re-adapt the GF to prevent the object from slipping [21]. In Parkinson's disease, the disruptions in the coordination of GF-LF can lead to imprecise grip and poor performance in lifting tasks [22].

8.3 Artificial sensing for fingertip, object, and environment contact

The human hands have been used as an inspiration source for the design of robotic hands. Robotic hands have been designed with similar number of fingers, sensing, and control algorithms for the grasping and manipulation of objects [9][8][23]. Analyses on dexterous manipulation begin with the object [24]. In other words, the dexterous manipulation problem is formulated in terms of the object to be manipulated, how it should behave, and what forces should be exerted upon it.

Successful dexterous manipulation requires control at the object-finger contact during manipulation tasks [25][26]. Because the interface between an object and the hand is the fingertip, the tactile sensor at the contact interface becomes extremely relevant. Tactile sensing allows the robot hand to perform adaptive control of the grasps and fingertip forces in relation to an object's physical properties, like texture and weight, and any disturbances like slippage or external pull [19]. In the next section, basic concepts in contact modeling and tactile sensor modeling are introduced to form the basis of the bioinspired tactile contact sensing.

8.3.1 Finger-object contact modeling

Prattichizzo and Trinkle [27] described three primary contact models for grasp analysis: Point Contact Without Friction (PwoF; Fig. 8.2a), Hard Finger (HF; Fig. 8.2b), and Soft Finger (SF; Fig. 8.2c). The PwoF model is applicable when the contact patch is very small and the surfaces are slippery, allowing only the normal component of the contact force to be transmitted, while neglecting tangential forces and moments. The HF model is used when there is significant contact friction but no appreciable friction moment, transmitting all three translational velocity components and contact forces, but not angular velocities or moments. In contrast, the SF model is employed when both surface friction and the contact patch are large enough to generate significant friction forces and moments, allowing for the transmission of all three translational velocity components, the angular velocity about the contact normal, and the normal component of the contact moment.

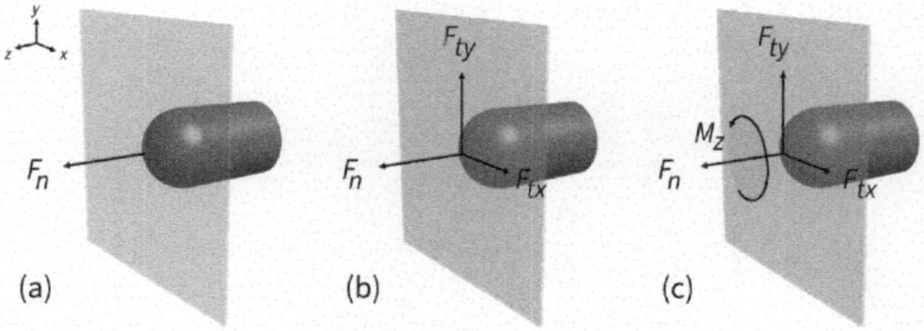

FIGURE 8.2 Typical contact models with their forces and moments: (a) frictionless point contact, (b) point contact with friction, (c) soft finger contact.

The case of a point contact without friction was earlier solved by Prattichizzo and Trinkle [27] in relation to a sphere which is grasped by a two-fingered robot hand with five joints. In Fig. 8.3a, consider that the sphere is three-dimensional with a radius, r, and it is shown to be in a planar view. The two fingers are in contact with the sphere at angles θ_1 and θ_2. The black dots represent the points of contact at the fingertip, where the frames $\{\boldsymbol{C}\}_1$ and $\{\boldsymbol{C}\}_2$ are positioned. The directions $\hat{o}_1$ and $\hat{o}_2$ extend perpendicularly from the plane of the figure. Frame $\{\boldsymbol{B}\}$ is positioned at the center of the sphere.

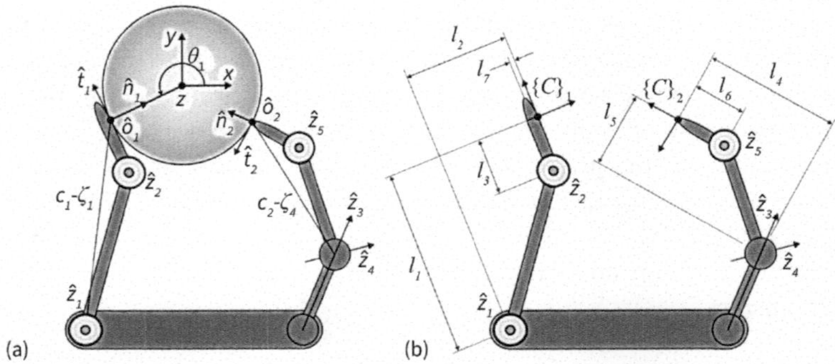

FIGURE 8.3 A two-fingered robot hand with five revolute joints. (a) The relevant variables when the robot hand is grasping a sphere. (b) The hand Jacobian data. Reproduced from Prattichizzo and Trinkle [27], with permission.

The z-axis direction is pointed outward from the page. The two joint axes of the left finger are orthogonal to the x and y plane. The finger on the right has three revolute joints. Its first and second axes $\hat{z_3}$ and $\hat{z_4}$ currently lie in the plane. Consequently, a rotation about $\hat{z_3}$ will result in $\hat{z_4}$ having an out-of-plane component. It would cause the fingertip of the right finger to leave the plane.

The rotation matrix for the i-th contact frame is given in [27] as:

$$R_i = \begin{pmatrix} -\cos(\theta_i) & \sin(\theta_i) & 0 \\ -\sin(\theta_i) & -\cos(\theta_i) & 0 \\ 0 & 0 & 1 \end{pmatrix} \tag{8.1}$$

The grasp matrix, $\tilde{\boldsymbol{G}}$, for contact i is given in [27] as:

$$\tilde{\boldsymbol{G}}_i = \left(\begin{array}{cccccc} -c_i & s_i & 0 & & & \\ -s_i & -c_i & 0 & & \mathbf{0} & \\ 0 & 0 & 1 & & & \\ \hline 0 & 0 & rs_i & -c_i & s_i & 0 \\ 0 & 0 & -rc_i & -s_i & -c_i & 0 \\ 0 & -r & 0 & 0 & 0 & 1 \end{array}\right) \tag{8.2}$$

where $\mathbf{0} \in \mathbb{R}^{3\times 3}$ is the zero matrix. c_i and s_i are the short forms for $\cos(\theta_i)$ and $\sin(\theta_i)$. The grasp matrix is defined as $\tilde{\boldsymbol{G}} = (\tilde{\boldsymbol{G}}_1\ \tilde{\boldsymbol{G}}_2) \in \mathbb{R}^{6\times 12}$.

Using the additional position information from Fig. 8.3b, the completed hand Jacobian can be constructed. The hand Jacobian, $\tilde{\boldsymbol{J}}_i$, for contact i requires the joint axis directions and the origins of the frames to be fixed to the links of each finger.

The complete hand Jacobian $\tilde{\boldsymbol{J}} \in \mathbb{R}^{12\times 5}$ is given in [27] as:

$$\tilde{\boldsymbol{J}} = \left(\begin{array}{ccccc} -l_1 & -l_3 & & & \\ l_2 & l_7 & & & \\ 0 & 0 & & & \\ 0 & 0 & & \mathbf{0} & \\ 0 & 0 & & & \\ 1 & 1 & & & \\ \hline & & 0 & 0 & 0 \\ & & 0 & 0 & l_6 \\ & & l_4 & \frac{\sqrt{2}}{2}(l_4 + l_5) & 0 \\ \mathbf{0} & & 0 & -\frac{\sqrt{2}}{2} & 0 \\ & & -1 & -\frac{\sqrt{2}}{2} & 0 \\ & & 0 & 0 & 1 \end{array}\right) \tag{8.3}$$

The horizontal partition denotes $\tilde{\boldsymbol{J}}_1$ on the top and $\tilde{\boldsymbol{J}}_2$ on the bottom. The columns on the top represent joints 1 and 2 of the left finger. The columns on the bottom represent joints 3, 4 and 5 of the right finger. It can be observed that the block diagonal structure resulted from the finger i directly affecting only contact i. Notice also from the hand Jacobian, $\tilde{\boldsymbol{J}}$, and in Fig. 8.3b that position information at the contact point is important. Various contact models for hard finger and soft finger contacts were further derived in [27].

8.3.2 Example: on-off contact sensors

8.3.2.1 Design objectives

In Rocella et al. [28] and in Edin et al. [11], on-off contact sensors were designed to emulate the mechanoreceptors of the human hand. Johansson and Vallbo [29] estimated that 90% of the Slowly Adapting type I (SA I) and the Fast Adapting type I (FA I) mechanoreceptors get excited to a stimulus of 5 mN. In terms of pressure, the 5 mN stimulus was applied through a von Frey hair with a diameter of 0.27 mm [29]. Therefore, contact activation below 87.33 mN/mm^2 represents the design goal for contact sensor development.

The on-off contact sensors (Fig. 8.4a) were developed using flexible circuits and were fabricated with the standard photolithography procedures on Kapton (Dupont Pyralux film LF9150R) [30]. The top and bottom Kapton sheets each have 127 μm thickness with a 35 μm layer of copper (305 g/m^2) on one side. Strips of polyurethane foam with a thickness of 2 mm were positioned between the two layers. The foam spacers allowed the top layer to return to its initial state upon the termination of contact with an object. Without these

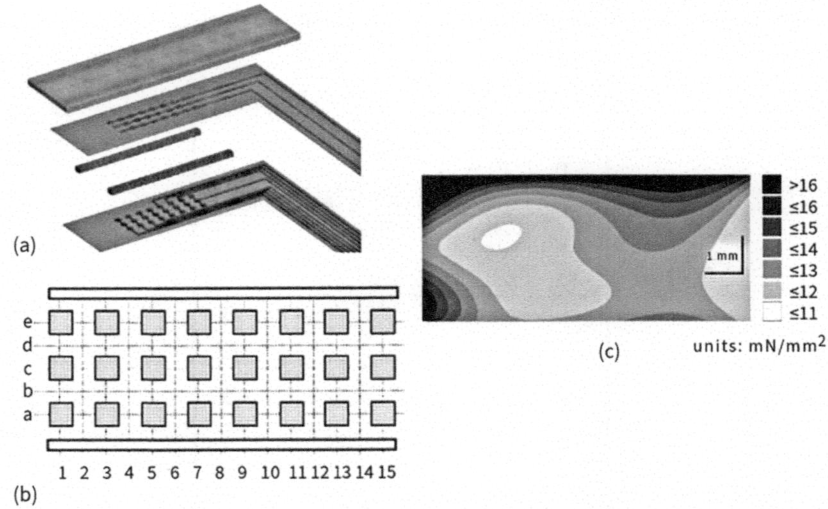

FIGURE 8.4 The contact sensor design and characterization results. (a) The contact sensor comprised two flexible layers with copper electrodes. Each electrode has a 1 mm × 1 mm area with a center-to-center distance of 1.4 mm. Two layers of foam support the 1 mm thick silicone skin layer. (b) The distribution and nomenclature for each array element, with a total of 24 tactile elements (taxels). (c) The contours of the tactile sensors' sensitivity to contact. The contours' color coding on the right side of the figure shows the minimum contact pressure to activate at least one of the taxels. The recorded threshold on the contact area $13.3 \pm 0.45 \times 10^{-3}$ N/mm^2 (mean ± S.D.). Reproduced from Edin et al. [11], with permission.

foam spacers, severe hysteresis was observed. Furthermore, this "unnecessary contact" became more evident when the layers were wrapped around the fingers.

8.3.2.2 Methodology

In the human hand, the highest density of tactile units can be found in the fingertip [31][29]. The on-off tactile sensor consisted of two long foam strips, an artificial skin (Fig. 8.4a), and a 3 × 8 array of tactile elements (Fig. 8.4b). In addition to the skin's ability to recruit more tactile elements, the skin can provide friction to increase grasp stability. The 1 mm thick silicone skin (GLS 40, Prochima s.n.c. Italy) was cast with a tube-like geometry and was intended to be slid over the contact sensor. The skin's inside diameter was designed to have a tight fit with the fingertip and contact sensor assembly so that the skin will not fall or slide off.

8.3.2.3 Results

Threshold estimations were obtained for the contact sensor array designed for the fingertip. The threshold recorded was less than 15 mN/mm^2. This result was way below the target activation value of 87.33 mN/mm^2. Fig. 8.4c shows the pressure distribution profile of the contact sensor.

As for the hysteresis, the difference between the pressure at which at least one contact sensor signals onset and subsequent offset when the pressure was released averaged less than 0.001 N/mm^2 ($7.3 \pm 3.8 \times 10^{-4}$ N/mm^2; mean±SD). These results imply that the hysteresis was not significant. In effect, the polyurethane foam spacers, which were positioned between the two contact layers, sprang back to their original state with very minimal loss during the compression (see [11] for more details).

8.3.3 Grasp and lift forces

Fig. 8.5 shows the free body diagram (FBD) of the various forces acting on a robot's fingers and a grasped object. A successful lift can be accomplished by coordinating two types of forces: the Grip Force (GF) and the Load Force (LF).

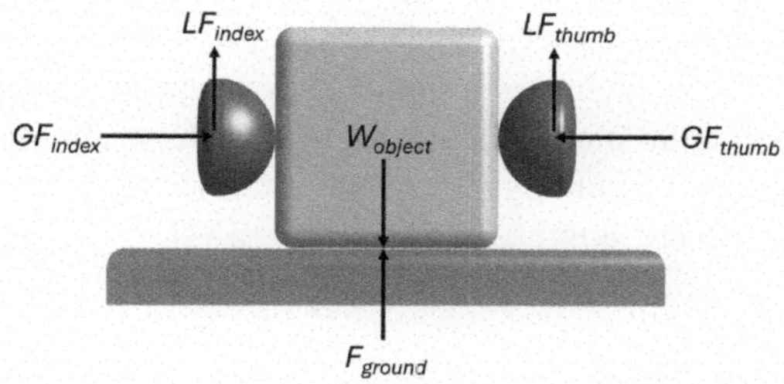

FIGURE 8.5 Forces involved in a fingertip grasping task. Free-body diagram (FBD) of a precision grasp showing the grip force (GF), load force (LF), weight of the object (W_{object}), and the ground reaction force (F_{ground}).

The Grip Force is shown as the horizontal force applied perpendicularly to the object's surfaces by the robot's fingers. These are the counteracting forces from the index finger GF_{index} and the thumb GF_{thumb}. The Load Force is the tangential force transmitted through the fingertip-object contacts as the robot arm lifts the object. Although this force is generated by the robot arm's actuators, the load force is transferred to the object through static friction at the contact interfaces.

The successful lifting of an object depends on the coordination between the grip force and the load force. The grip force generates sufficient static friction at the fingertip-object interfaces to transmit the load force without slippage. Increasing the grip force increases the maximum tangential force that can be transmitted to the object. Once the upward load force is greater than the weight of the object, the object is then lifted from the supporting surface.

The object is lifted when the upward load force exceeds the downward weight of the object. The grip force should be sufficiently large to prevent slip while avoiding excessive contact forces that could damage the object.

8.3.4 Example: triaxial force sensors

8.3.4.1 Design objectives

Again in Rocella et al. [28] and in Edin et al. [11], the triaxial force sensor in Fig. 8.6a was designed to satisfy the requirement of detecting the contact between the object and environment. Particularly, the goal was to evaluate whether the triaxial force sensor can detect the start and the termination of an object's vertical movement when lifted by the artificial hand. Inspired by the Sequential Coordination mechanism described in Sec. 8.2.1, the start of the object's vertical movement can give information to the control system to stop increasing the grip and lifting force being applied. The termination of the vertical movement can then send the cue to the controls system to unload. Just like in the human tactile system [10], these triggers may possibly help the controller when the weight of the object or the exact position of the support cannot be accurately predicted.

To be applicable, the triaxial force sensors should have sufficient bandwidth that would correspond to those of the SA and FA mechanoreceptors ranging from 2-400 Hz [32]. As such, the designed sensor was not intended to directly correlate to any particular mechanoreceptor type. Indeed, all types of biological mechanoreceptors are essentially force sensors since they are all capable of detecting forces when the skin is compressed [33][34].

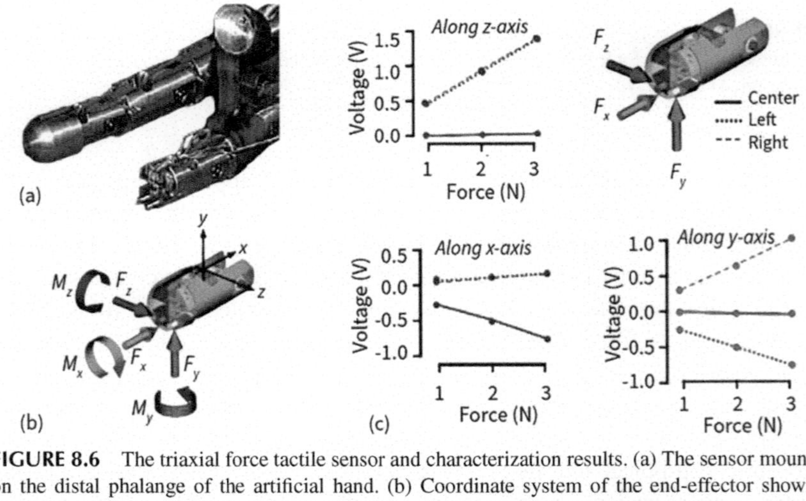

FIGURE 8.6 The triaxial force tactile sensor and characterization results. (a) The sensor mounted on the distal phalange of the artificial hand. (b) Coordinate system of the end-effector showing the three translational axes (F_x, F_y, F_z) and the three rotational components (M_x, M_y, M_z). (c) Characterization of the response of the triaxial force sensor. The sensor exhibited linear responses to forces applied independently along the x-, y-, and z-axes. Reproduced from Edin et al. [11], with permission.

8.3.4.2 Methodology

The force sensor was designed using finite element analysis software to identify the maximum strain levels and to select the appropriate commercial strain gauges. The actual sensor mounted at the fingertips was based on a flexible aluminum alloy structure. In order to detect the three components of an applied force, three metal-based strain gauges (N3K-06-S022H-50C, Vishay Micro-Measurements, Vishay Intertechnology Inc., PA) were attached to the root of each tether along the three axes of the sensor itself (Fig. 8.6b). Three additional strain gauges were used for temperature compensation. The sensor was integrated at the tip of the distal phalanx.

8.3.4.3 Results

The raw outputs from all strain gauges during static characterization were linearly related to the loads (Fig. 8.6c). Due to the mechanical structure of the sensor, it was possible to obtain a quantitative estimation of F_x, F_y and F_z components of the generic applied load through an appropriate combination of output from the strain gauges. The calibration curves show that the detection of stimuli along the x-direction relied only on the central strain gauge, while to measure stimuli along the y and z directions, combinations of the left and right strain gauges signals had to be computed.

8.4 Case study: object grasp-lift-return task

8.4.1 Methodology

In Edin et al. [35], the grasp-lift-return task was done with the Cyberhand (Fig. 8.7a). Together with a robot arm, this artificial hand was programmed to perform a pinch grasp on three types of plastic boxes with weights of 100, 200, 400 g. For the on-off contact sensors and the triaxial force sensors, the signals were acquired and sampled at the rate of 1 kHz and stored for post-processing. The contact between the object and the table was obtained from the first derivative of the strain gauge of the triaxial force sensor. A low-pass filter was applied to those signals.

8.4.2 Results

The phases in the grasp-lift-return task by the human hand (cf. Sec. 8.2.1; Fig. 8.7c) were again shown for comparison with the results from the same task by an artificial hand (Fig. 8.7d). The step response output from the on-off contact sensor showed the bursts during static contact of the SA I type mechanoreceptor. The filtered first derivative result from the triaxial force sensor showed the burst in FA II type mechanoreceptor when the object was lifted from the table and then returned to the table.

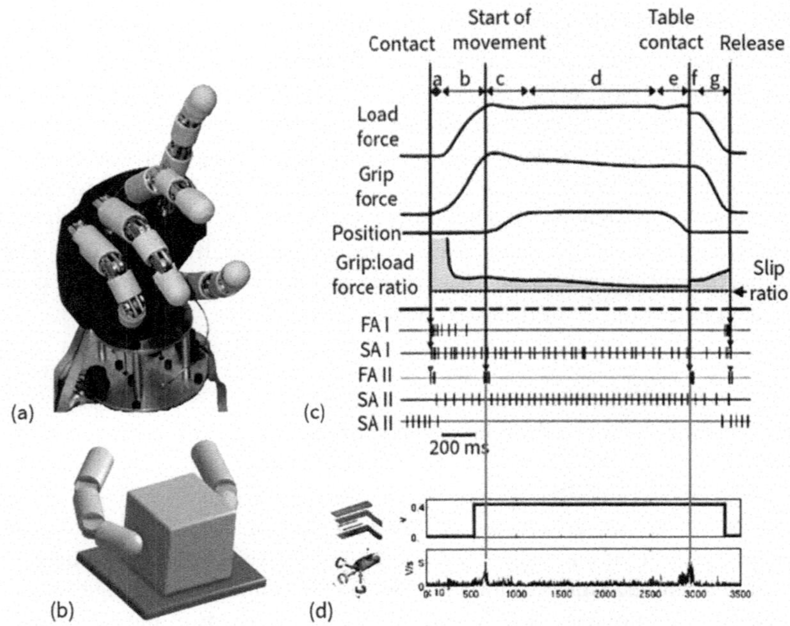

FIGURE 8.7 Object grasp, lift, and return task with an artificial hand. (a) The artificial hand, Cyberhand, used in this study composed of five fingers with cylindrical phalanges of aluminum alloy material attached to a central palm. It has embedded contact and triaxial force sensors. Reproduced from Edin et al. [35], with permission. ©2006 IEEE. (b) Schematic illustration of a grasp, lift, and return task. (c) The phases of a grasp, lift, and return task from the human thumb and index fingers. Reproduced from Edin et al. [11], with permission. ©2008 Elsevier. All rights reserved. (d) The response of the contact sensor showing the duration of contact and the spikes in the triaxial force sensor indicating the instant when the object has been lifted from and returned to the table (see red lines). Adapted from Edin et al. [35], with permission. ©2006 IEEE. All rights reserved.

8.5 Summary

The majority of the hands' and fingers' interactions with the world involve picking an object and holding it in a stable manner. Understanding the basic sensing requirements of such tasks is crucial. These tasks involve sensing the contact between the fingertip and the object, which then sends confirmation to start the next task (i.e., to lift an object from the environment).

The developments in the field of tactile sensors have been focused on the individual sensor components rather than as a complete system for tactile sensing. Thus, a two-fold approach is proposed in developing a bioinspired tactile sensory system. The first is at the component level while the other is at the system level. At a component level, the human mechanoreceptors provide some guidance. With this as the model, it is not necessary to fabricate sensors that have extremely accurate quantitative output. Bioinspired tactile sensors should be regarded as devices that extract qualitative or semi-quantitative information rather than as ultraprecision devices [36][37][38]. The sensor components must be developed within the bounds of currently available technologies but have functional resemblance (i.e., time/frequency response, threshold) with their biological counterpart.

At the system level, the sensor components are proposed to have a response, which is comparable with the response of the human mechanoreceptors in grasping and manipulation tasks. The fingertip pinch grasp on the object and lifting the object from a table is a basic human activity that has been extensively studied in neurophysiology. From this, the requirements of the artificial tactile sensory system have been extracted. The phase transitions that occur during the object lifting task would be useful to enable the artificial hand controller to distinguish between the four main phases of the task (i.e., contact and pre-load, lifting, contact with the table, releasing) in order to implement a bioinspired control of the hand.

The chapter describes a bioinspired tactile sensing system designed for artificial hands, focusing on the critical interactions at the finger-object and object-environment interfaces. The development and evaluation of on-off contact sensors and triaxial force sensors are some of the potential solutions to replicate some of the sensory capabilities of the human hand. The on-off contact sensors demonstrated remarkable sensitivity. It achieved pressure thresholds below 15 mN/mm^2, which closely align with the activation thresholds of human Slowly Adapting type I (SA I) mechanoreceptors [29].

The triaxial force sensors can provide force measurements along three axes, allowing for the detection of dynamic mechanical events during object manipulation. In Edin et al. [11], the sensors were able to exhibit a frequency response and mechanical resonance consistent with theoretical prediction. The resonance frequency was around 2.7 kHz and an additional peak near 389 Hz, to capture rapid force changes in the human fingertips [10][19].

The feedback from the contact and force sensors is essential information to the control of artificial hands to hold an object in a stable manner and manipulate it, similar to how we operate on objects we interact with on a daily basis. The bioinspired tactile system described offers a promising pathway for enhancing robotic manipulation capabilities [29][1][27].

References

[1] P. Dario, D. De Rossi, Tactile Sensors and the Gripping Challenge, IEEE Spectrum, 1985, pp. 46–52.

[2] M. Lee, Nicholls H. Review, Article Tactile sensing for mechatronics—a state of the art survey, Mechatronics 9 (1999 Feb) 1–31, https://doi.org/10.1016/s0957-4158(98)00045-2.

[3] J. Dargahi, S. Najarian, Human tactile perception as a standard for artificial tactile sensing—a review, The International Journal of Medical Robotics and Computer Assisted Surgery 1 (2004) 23–35, https://doi.org/10.1002/rcs.3.

[4] R. Dahiya, G. Metta, M. Valle, G. Sandini, Tactile sensing—from humans to humanoids, IEEE Transactions on Robotics 26 (2010 Feb) 1–20, https://doi.org/10.1109/tro.2009.2033627.

[5] S. Jacobsen, I. McCammon, K. Biggers, R. Phillips, Tactile Sensing System Design Issues in Machine Manipulation, Proceedings. 1987 IEEE International Conference on Robotics and Automation, vol. 4, Institute of Electrical and Electronics Engineers, 1987, pp. 2087–2096, https://doi.org/10.1109/robot.1987.1087755.

[6] R. Johansson, G. Westling, Signals in tactile afferents from the fingers eliciting adaptive motor responses during precision grip, Experimental Brain Research 66 (1987 Mar), https://doi.org/10.1007/bf00236210.

[7] R. Johansson, B. Edin, Predictive feed-forward sensory control during grasping and manipulation in man, Biomedical Research 14 (1993 Jan).

[8] M.T. Mason, J.K. Salisbury Jr., Robot Hands and the Mechanics of Manipulation, MIT Press, Cambridge, MA, 1985.

[9] M. Cutkosky, On grasp choice, grasp models, and the design of hands for manufacturing tasks, IEEE Transactions on Robotics and Automation 5 (1989 Jun) 269–279, https://doi.org/10.1109/70.34763.

[10] G. Westling, R.S. Johansson, Responses in glabrous skin mechanoreceptors during precision grip in humans, Experimental Brain Research 66 (1987) 128–140, https://doi.org/10.1007/BF00236209.

[11] B. Edin, L. Ascari, L. Beccai, S. Roccella, J.J. Cabibihan, M. Carrozza, Bio-inspired sensorization of a biomechatronic robot hand for the grasp-and-lift task, Brain Research Bulletin 75 (2008) 785–795, https://doi.org/10.1016/j.brainresbull.2008.01.017.

[12] R.S. Johansson, G. Westling, Tactile afferent signals in the control of precision grip, in: M. Jeannerod (Ed.), Attention and Performance XIII: Motor Representation and Control, Lawrence Erlbaum Associates, Hillsdale, NJ, 1990, pp. 677–713.

[13] B.B. Edin, G. Westling, R.S. Johansson, Independent control of human finger-tip forces at individual digits during precision lifting, Journal of Physiology 450 (1992) 547–564, https://doi.org/10.1113/jphysiol.1992.sp019142.

[14] R.S. Johansson, G. Westling, Roles of glabrous skin receptors and sensorimotor memory in automatic control of precision grip when lifting rougher or more slippery objects, Experimental Brain Research 56 (1984) 550–564, https://doi.org/10.1007/bf00237997.

[15] R.S. Johansson, G. Westling, Afferent signals during manipulative tasks in humans, in: O. Franzén, J. Westman (Eds.), Information Processing in the Somatosensory System, in: Wenner-Gren Center International Symposium Series, Palgrave (Macmillan), London, 1991, pp. 25–48, https://doi.org/10.1007/978-1-349-11597-6_3.

[16] H.H. Ehrsson, E. Fagergren, H. Forssberg, Differential fronto-parietal activation depending on force used in a precision grip task: an fMRI study, Journal of Neurophysiology 85 (2001) 2613–2623, https://doi.org/10.1152/jn.2001.85.6.2613.

[17] H.H. Ehrsson, A. Fagergren, R.S. Johansson, H. Forssberg, Evidence for the involvement of the posterior parietal cortex in coordination of fingertip forces for grasp stability in manipulation, Journal of Neurophysiology 90 (2003) 2978–2986, https://doi.org/10.1152/jn.00958.2002.

[18] J.R. Flanagan, A.M. Wing, Modulation of grip force with load force during point-to-point arm movements, Experimental Brain Research 95 (1993) 131–143, https://doi.org/10.1007/BF00229662.

[19] R.S. Johansson, J.R. Flanagan, Coding and use of tactile signals from the fingertips in object manipulation tasks, Nature Reviews. Neuroscience 10 (2009 Apr) 345–359, https://doi.org/10.1038/nrn2621.

[20] J. Zhang, Y. Xiao, Z.M. Li, N. Wei, L. Lin, K. Li, Reach-to-grasp kinematics and kinetics with and without visual feedback in early-stage Alzheimer's disease, Journal of NeuroEngineering and Rehabilitation 19 (2022), https://doi.org/10.1186/s12984-022-01108-1.

[21] A. Zangrandi, M. D'Alonzo, C. Cipriani, G. Di Pino, Neurophysiology of slip sensation and grip reaction: insights for hand prosthesis control of slippage, Journal of Neurophysiology 126 (2021) 477–492, https://doi.org/10.1152/jn.00087.2021.

[22] P.E. Ingvarsson, A.M. Gordon, H. Forssberg, Coordination of manipulative forces in Parkinson's disease, Experimental Neurology 145 (1997) 489–501, https://doi.org/10.1006/exnr.1997.6480.
[23] R.D. Howe, Tactile sensing and control of robotic manipulation, Advanced Robotics 8 (1993 Jan) 245–261, https://doi.org/10.1163/156855394x00356.
[24] A.M. Okamura, N. Smaby, M.R. Cutkosky, An overview of dexterous manipulation, in: Proceedings of the 2000 IEEE International Conference on Robotics and Automation (ICRA), San Francisco, CA, 2000, pp. 255–262, https://doi.org/10.1109/ROBOT.2000.844067.
[25] R.S. Fearing, Tactile sensing mechanisms, The International Journal of Robotics Research 9 (1990) 3–23, https://doi.org/10.1177/027836499000900301.
[26] H. Zhang, N. Chen, Control of contact via tactile sensing, IEEE Transactions on Robotics and Automation 16 (2000) 482–495, https://doi.org/10.1109/70.880799.
[27] D. Prattichizzo, Trinkle JC. Grasping, in: B. Siciliano, O. Khatib (Eds.), Springer Handbook of Robotics, Springer, Berlin, Heidelberg, 2008, pp. 671–700, https://doi.org/10.1007/978-3-540-30301-5_29.
[28] S. Roccella, M.C. Carrozza, G. Cappiello, J.J. Cabibihan, C. Laschi, P. Dario, H. Takanobu, M. Matsumoto, H. Miwa, K. Itoh, A. Takanishi, Design and development of five-fingered hands for a humanoid emotion expression robot, International Journal of Humanoid Robotics 04 (2007 Mar) 181–206, https://doi.org/10.1142/s0219843607000996.
[29] R.S. Johansson, A.B. Vallbo, Spatial properties of the population of mechanoreceptive units in the glabrous skin of the human hand, Brain Research 184 (1980) 353–366, https://doi.org/10.1016/0006-8993(80)90804-5.
[30] M. Zecca, S. Roccella, M.C. Carrozza, H. Miwa, K. Itoh, G. Cappiello, J.J. Cabibihan, M. Matsumoto, H. Takanobu, P. Dario, A. Takanishi, On the development of the emotion expression humanoid robot WE-4RII with RCH-1, in: Proceedings of the 4th IEEE/RAS International Conference on Humanoid Robots, 2004, pp. 235–252, https://doi.org/10.1109/ICHR.2004.1442125.
[31] R. Johansson, A. Vallbo, Tactile sensibility in the human hand: relative and absolute densities of four types of mechanoreceptive units in glabrous skin, Journal of Physiology 286 (1979) 283–300, https://doi.org/10.1113/jphysiol.1979.sp012619.
[32] R.S. Johansson, U. Landström, R. Lundström, Responses of mechanoreceptive afferent units in the glabrous skin of the human hand to sinusoidal skin displacements, Brain Research 244 (1982) 17–25, https://doi.org/10.1016/0006-8993(82)90899-X.
[33] I. Birznieks, P. Jenmalm, A.W. Goodwin, R.S. Johansson, Encoding of direction of fingertip forces by human tactile afferents, Journal of Neuroscience 21 (2001) 8222–8237, https://doi.org/10.1523/jneurosci.21-20-08222.2001.
[34] R.S. Johansson, I. Birznieks, First spikes in ensembles of human tactile afferents code complex spatial fingertip events, Nature Neuroscience 7 (2004 Jan) 170–177, https://doi.org/10.1038/nn1177.
[35] B. Edin, L. Beccai, L. Ascari, S. Roccella, J. Cabibihan, M. Carrozza, Bio-inspired approach for the design and characterization of a tactile sensory system for a cybernetic prosthetic hand, in: Proceedings 2006 IEEE International Conference on Robotics and Automation, ICRA 2006, 2006, pp. 1354–1358, https://doi.org/10.1109/robot.2006.1641897.
[36] R. Lazzarini, R. Magni, P. Dario, A tactile array sensor layered in an artificial skin, in: Proceedings of the 1995 IEEE/RSJ International Conference on Intelligent Robots and Systems (IROS'95), Pittsburgh, PA, USA, vol. 3, 1995, pp. 114–119, https://doi.org/10.1109/IROS.1995.525871.
[37] P. Dario, R. Lazzarini, R. Magni, S.R. Oh, An integrated miniature fingertip sensor, in: Proceedings of the 7th International Symposium on Micro Machine and Human Science (MHS'96), Nagoya, Japan, 1996, pp. 91–97.
[38] D. Taddeucci, C. Laschi, R. Lazzarini, R. Magni, P. Dario, A. Starita, An approach to integrated tactile perception, in: Proceedings of the 1997 IEEE International Conference on Robotics and Automation (ICRA), Albuquerque, NM, USA, vol. 4, 1997, pp. 3100–3105, https://doi.org/10.1109/ROBOT.1997.606759.

Chapter 9

Sensing object slippage and reflex response

9.1 Introduction

Successful dexterous manipulation requires that the slippage of an object is effectively sensed and controlled. This task is made possible by the complex operation of the human tactile sensing system. The interaction between the object and the human fingertips is monitored by the mechanoreceptors at the fingertip. These mechanoreceptors, comprising the Meissner corpuscles, Pacinian corpuscles, Merkel cells, and Ruffini endings, are located within the layers of the fingertip, with each one serving a unique function [1][2]. Each mechanoreceptor has its particular sensitivity for pressure, texture, and vibration. The interaction between these detectors and the processing function of the central nervous system ensures that objects are held securely even with dynamically changing conditions [3][4][5].

Slip detection in humans manifests in a two-phase pattern. First, localized slip events occur at the peripheral zones of a contact region. This is termed as incipient slippage. Next, as the tangential stress increases, a complete slippage phase occurs where the whole contact area undergoes slippage [6][7][8]. Humans are able to sense incipient slippage such that the complete slippage is controlled through the adjustment of the grip force [9]. It was reported that the central nervous system's capability to interpret slippage information is important for controlling a grasped object in a highly dynamic condition when disturbances may occur [10].

This chapter investigates the principles of slip detection in humans with the aim of replicating the slip capability for robotic or prosthetic hands. The chapter examines the dynamics of the finger contact with the object, when the object has been lifted from the table. An optical sensor and the slip detection algorithm were described based on the work of Nakagawa et al. [11]. This system provided the basis for a reflex response system that monitors the minor movements on a grasped object and responds swiftly to any external disturbances.

Sec. 9.2 presents a brief background of the human mechanism for slip sensing where the roles of the mechanoreceptors are described. Sec. 9.3 discusses the contact mechanics and the control system block diagram for a grasped object by artificial fingers. Sec. 9.4 describes the design and evaluation of optical tactile sensing for object slippage. Sec. 9.5 demonstrates

Humanizing Artificial Touch. https://doi.org/10.1016/B978-0-32-395519-5.00020-9

the implementation of the DATH algorithm that was implemented on an artificial hand during the grasping and control of dark and transparent drinking bottles. The last section summarizes the chapter.

9.2 Human physiology of slippage sensing

9.2.1 Mechanoreceptors

Human dexterous manipulation functions through a synchronized coordination of specialized skin mechanoreceptors, feedback system, and the integration of sensory information by the central nervous system. Various types of mechanoreceptors are required to accomplish this task. As mentioned in Chap. 7, there were four types of mechanoreceptors that were described. These were the Meissner corpuscles (FA I), Pacinian corpuscles (FA II), Merkel cells (SA I), and Ruffini endings (SA II).

The Fast Adapting Type 1 and Type 2 mechanoreceptors are mainly responsible for slippage sensing. Located near the fingertip surface are the Meissner corpuscles (FA I), which are more sensitive to texture and vibrations. Positioned deeper from the surface are the Pacinian corpuscles, which respond to high-frequency vibrations and rapid changes in pressure [12]. Subtle changes in the object-skin contact are possible due to the sensitivity of these receptors (cf. Fig. 7.1).

9.2.2 Incipient slippage

Just before an object begins to slip, localized slip events occur at the periphery of the contact area between the object and the finger. The deformations in the fingertip skin are detected by the mechanoreceptors [6][7][8]. Barea et al. [9] showed that humans can perceive slip before it becomes a full slip. Their studies indicate that the experiment participants can detect slippage when only a portion of the contact area is slipping. The early detection of the slip events permits the adjustment of the grip force to maintain the stability of the grasp.

The information from mechanoreceptors is processed by the CNS to modulate the grip force in real-time. The human tactile system ensures that the applied force is sufficient to balance the forces, such as the object's weight and inertia, that are acting on the object [1]. Due to this real-time monitoring and processing of the mechanoreceptor signals, it has also been reported that the applied grip force adapts to the frictional properties of the object being grasped [3][2]. As shown in Fig. 9.1, experiment participants tend to apply greater grip force to prevent slippage, while less force is required for objects with higher friction [3].

The timing of the responses from the tactile afferent has been found to be equally important for the fast encoding and transmission of the slip-related information to the CNS. In Delhaye et al. [10], they showed that the relative timing from different mechanoreceptors' impulses offers useful cues related to

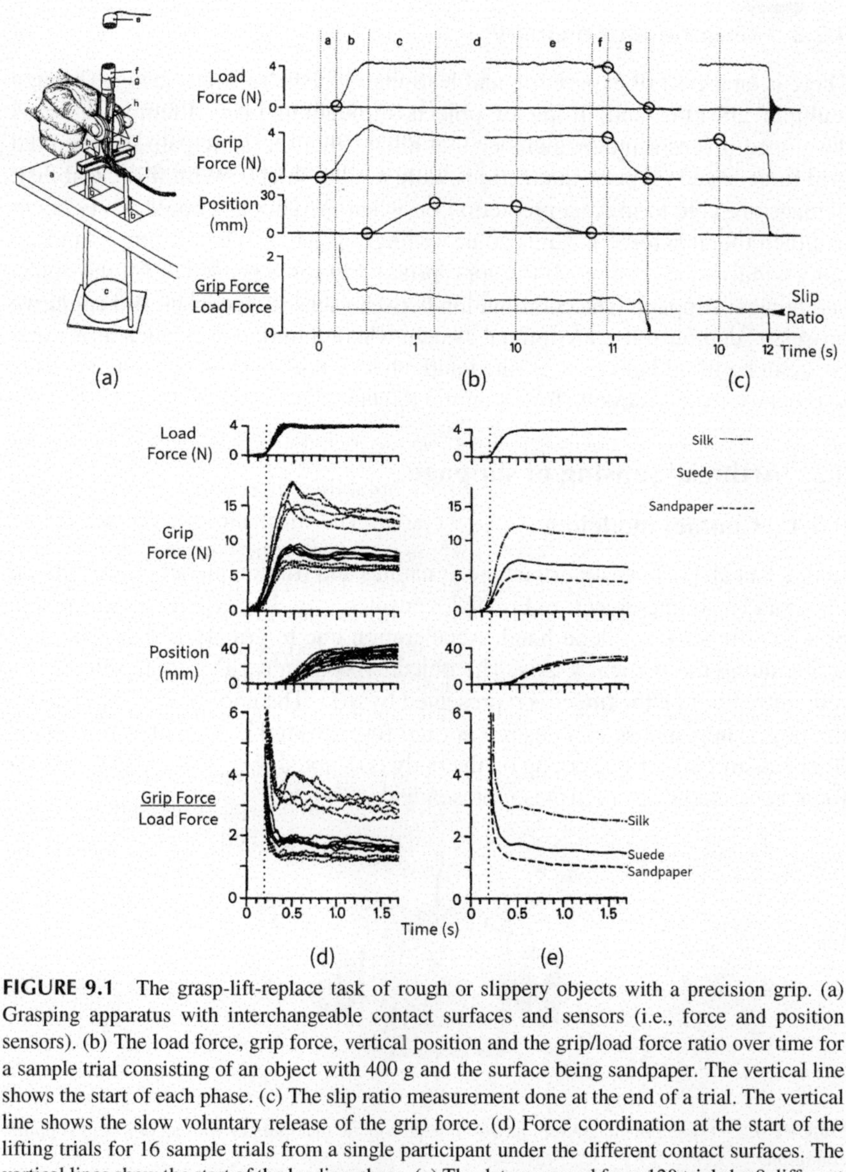

FIGURE 9.1 The grasp-lift-replace task of rough or slippery objects with a precision grip. (a) Grasping apparatus with interchangeable contact surfaces and sensors (i.e., force and position sensors). (b) The load force, grip force, vertical position and the grip/load force ratio over time for a sample trial consisting of an object with 400 g and the surface being sandpaper. The vertical line shows the start of each phase. (c) The slip ratio measurement done at the end of a trial. The vertical line shows the slow voluntary release of the grip force. (d) Force coordination at the start of the lifting trials for 16 sample trials from a single participant under the different contact surfaces. The vertical lines show the start of the loading phase. (e) The data averaged from 120 trials by 9 different participants. Reproduced from Johansson and Westling [3], with permission. ©1984 Springer. All rights reserved.

the slip onset. This results in the adjustments in the grip force. For unexpected changes in load or surface conditions, such a mechanism for fast adjustments is critical to prevent objects being dropped by the fingers.

9.2.3 Sensorimotor memory

There is an aspect of experience and learning involved in slip sensing. Through multiple grasping and lifting of objects throughout one's lifetime, humans developed a sensorimotor memory that allows them to anticipate the required grip force based on prior interactions with similar objects. With this capability, humans are able to make a prediction on when the slippage could likely occur and then the grip force is adjusted accordingly [3][13]. When lifting a familiar object, individuals can recall the appropriate grip force needed to lift the object and prevent slippage. The coordination between sensory feedback and cognitive processes demonstrates an optimal system where a light object, such as a paper cup, can be lifted but is not crushed and a heavy book can be lifted and carried without exerting excessive forces on the fingers.

9.3 Artificial sensing of slippage

9.3.1 Contact modeling

In Prach et al. [14], we described the dynamics and friction model of an artificial finger and a sliding object. In Fig. 9.2, consider a rigid body with mass, m, held by a two-fingered artificial hand. Acceleration due to gravity is denoted by g, acting along the vertical y-axis. The object's position relative to an initial zero reference point on the finger is represented by $q(t)$. The normal force applied by the finger, in response to a desired normal force from the controller, is $F_n(t) \geq 0$. The disturbance force acting on the body is denoted by $F_d(t)$, and the friction force generated due to $F_n(t)$ is represented by $F_f(t)$.

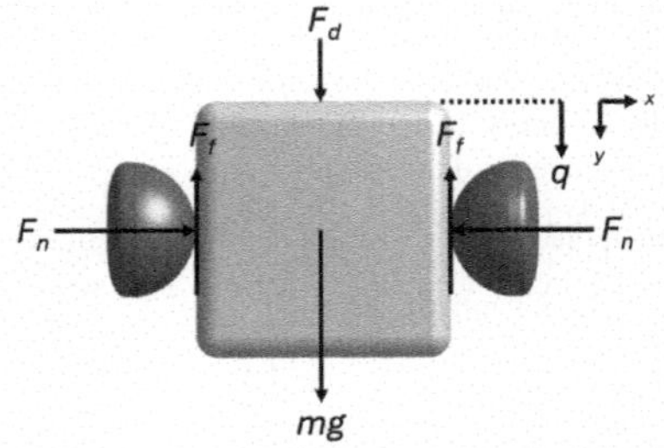

FIGURE 9.2 Free body diagram of a two-fingered robotic manipulator with an object in slipping condition.

When the object's velocity is zero, $\dot{q} = 0$, the object is fixed relative to the fingers, indicating no slipping. Conversely, if $\dot{q} \neq 0$, the object is slipping relative to the fingers. This slipping can occur in either direction, as the disturbance force, $F_d(t)$, can be positive or negative. The equation of motion governing the body's movement is given by:

$$mg + F_d(t) - F_f(t) = m\ddot{q}(t) \tag{9.1}$$

where $F_f(t)$ is defined by the friction model used. Expressing this equation in terms of the body's velocity, $v(t) \triangleq \dot{q}(t)$, Eq. (9.1) leads to:

$$mg + F_d(t) - F_f(t) = m\dot{v}(t). \tag{9.2}$$

The friction force is crucial for determining whether the object will slip or remain in place. If the velocity $v(t)$ is determined to be zero, the object is considered to be in a state of static equilibrium. In such a case, it can be said that there was no slippage that occurred. However, if $v(t) \neq 0$, the object is slipping relative to the artificial finger. In this case, the control system would have to adjust the grip force that is applied to the object.

To quantify how far the object has moved relative to its initial position, the slip distance $\delta(t)$, is defined as:

$$\delta(t) \triangleq q(t) - q(0) = \int_0^t v(\tau)d\tau - q(0) \geq 0. \tag{9.3}$$

The asymptotic slip distance is the distance over which the slip behavior approaches a steady state and is given by:

$$\delta_\infty \triangleq \lim_{t\to\infty} \delta(t). \tag{9.4}$$

The friction force, F_f, is modeled using the Coulomb friction model [15], which can be expressed as:

$$F_f = \begin{cases} F_C sgn(v), & v \neq 0; \\ min(|F_d|, F_{st}) sgn(F_d), & v = 0, \end{cases} \tag{9.5}$$

where F_C denotes the Coulomb friction force and F_{st} denotes static friction force. These are further defined as:

$$F_C = \mu F_n, \tag{9.6}$$

$$F_{st} = \mu_{st} F_n, \tag{9.7}$$

where $\mu > 0$ is the sliding friction coefficient and $\mu_{st} > 0$ is the static friction coefficient. In the stick-slip model, the stick condition happens when $v = 0$. The direction of the friction force during static conditions is determined by the direction of the disturbance force, F_d. The friction force becomes proportional to the applied normal force by the artificial finger when an object is slipping.

The dynamics of the finger-object system are influenced by various factors. These factors include the mass of the object, the coefficients of friction, and the applied forces. The control system must be designed to respond to changes in these parameters to successfully prevent the object's slippage. The objective of the control system is to regulate the normal force, F_n, that the finger is applying to the object so that the slip velocity, $v(t)$, approaches zero while minimizing the slip distance, $\delta(t)$.

9.3.2 Slippage control

Fig. 9.3 describes the feedback control architecture proposed in Prach et al. [14]. In this architecture, the controller was intended to regulate the normal force, F_n, applied by the robotic manipulator (see Fig. 9.2). The objective is to get the object's slip velocity, v, to a stop and with the slip distance, $\delta(t)$, to be at a minimum. Moreover, the artificial finger needs to regulate the applied force without excessive grasping force to prevent damage to the object being grasped.

With the reference command to the control system being $v_{ref} = 0$, the feedforward loop of the architecture, as proposed in Prach et al. [14], would need a linear or nonlinear proportional-integral (PI) controller. The friction law is given by Engeberg et al. [16]. The disturbance force, F_d, which is applied to the body and the force on the body due to gravity, mg, are accounted for. The feedback loop accounts for the sensor noise, w, which was assumed to be Gaussian white noise. The velocity error, e, for the slippage is defined as $e(t) = v_{ref}(t) - v(t) = -v(t)$.

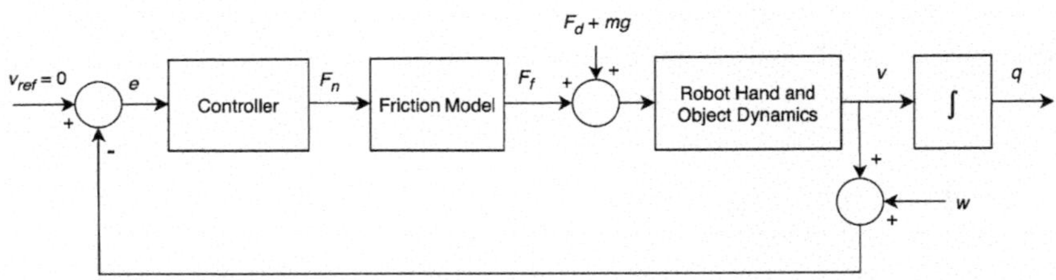

FIGURE 9.3 Control system block diagram for a robotic manipulator to suppress slippage of an object.

9.4 Example: optical tactile sensing for object slippage

Nakagawa-Silva et al. [17][11] presented the design and characterization of an optical slip sensor intended for its integration into artificial hands. The design objectives for this slippage sensor were as follows:

1. Miniaturize the sensor to enable integration into the fingers of dexterous artificial fingers.
2. Find or develop a sensor with high spatial resolution to detect the subtle motions associated with incipient and ongoing slip.
3. Consider a sensor that is robust to variations in object surface properties, including the reflectance of objects.
4. Make the system compatible with neuromorphic processing techniques to enable reflex-like responses to slippage events.

9.4.1 Slippage sensor

The slippage sensor (Fig. 9.4) is a miniature reflective optical sensor (OPR 5005, TT Electronics, USA). It integrates an infrared GaAIAs LED and a silicon phototransistor into a compact 3 × 4 × 1 mm^3 package. The form factor makes it suitable for integration into the fingers of artificial hands. The sensor's operating principle is based on detecting the changes in the amount of reflected infrared light caused by relative motion between the sensor and the surface of the grasped object.

The sliding of an object across the sensor's field of view causes a modulation of the intensity of the reflected light from the object's surface texture. The sensor's analog output voltage fluctuates as a result. These fluctuations are then processed using an event-based slip detection algorithm inspired by the behavior of Fast Adapting Type I (FA I) mechanoreceptors in the human skin.

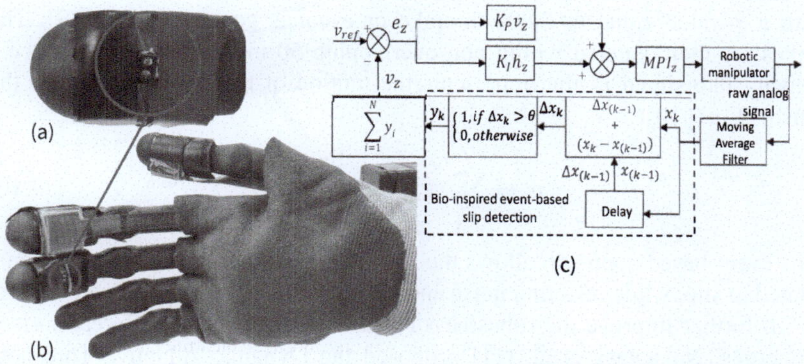

FIGURE 9.4 The slippage sensor at the fingertip of a robot hand and the control system architecture for slip suppression. (a) The slippage sensor positioned at the distal phalanx of the middle finger of the artificial hand. (b) The artificial hand i-Limb, which was used for the slippage experiments. (c) The proposed slippage control architecture, which combined a bio-inspired slip detection method with the Monotonic Proportional-Integral (MPI) control law to prevent object slippage. The raw analog signal is first filtered, and every sample, x_k, was used to generate spikes. The number of spikes, v_z, in a given time window is the feedback signal to the controller. The error signal, e_z, was given by the absolute difference between v_{ref} and v_z. The MPI control law controls the finger positions of the i-Limb and responds to slip by increasing the closure of the fingers and the grip force. Adapted from Nakagawa et al. [11], with permission. ©2019 IEEE.

9.4.2 Algorithm

Nakagawa-Silva et al. [17][11] proposed a neuromorphic event-based detection method for slippage control. In this method, the raw analog voltage from the slip sensor was low-pass filtered at 10 Hz. Then, the difference between consecutive filtered samples was accumulated over time:

$$\Delta x_k = \Delta x_{(k-1)} + (x_k - x_{(k-1)}), \quad k \geq 1 \tag{9.8}$$

where x_k is the current sample of the sensor signal, $x_{(k-1)}$ is the previous sample, and Δx_k is the current accumulation value.

The absolute value of x_k was then evaluated against a threshold θ. When x_k surpasses θ, an event or spike ($y_k = 1$) is produced. Upon the generation of an event, the value of x_k is reinitialized to start a new integration process. The slip detection methodology is defined as:

$$y_k = \begin{cases} 1, & if|\Delta x_k| > \theta; \\ 0, & otherwise, \end{cases} \tag{9.9}$$

where y_k represents the result of the event detection mechanism, whereas θ denotes the threshold that establishes the criteria for triggering an event.

When the absolute value of this accumulated difference crosses a threshold with a θ value equal to 0.008, a spike or event is generated (Eq. 9.9). The spikes are then counted within non-overlapping 50 ms windows to produce a velocity-like signal v_z that represents the motion of the object relative to the sensor (Eq. 9.10):

$$v_z = \sum_{i=1}^{N} y_{i,N} \quad (N \leq 16). \tag{9.10}$$

This event-based approach allows the sensor to highlight changes in the signal related to slip, while rejecting noise and fluctuations unrelated to object motion.

To further improve the robustness of the slip detection to different surface properties, Nakagawa et al. [11] proposed the Dynamic Adaptive Threshold (DATH) method. DATH is activated whenever a slip event is first detected. It then temporarily adjusts the threshold, θ, to be proportional to the instantaneous difference between the current and previous filtered sensor samples (Eq. (9.11)).

The threshold is stated as follows:

$$\theta_{DATH} = |x_k - x_{(k-1)}| \tag{9.11}$$

where the current sample is denoted as x_k and the previous sample is denoted as $x_{(k-1)}$. This adaptive thresholding strategy amplifies the sensitivity of the event detector for surfaces with lower reflectance (i.e., dark or transparent objects).

9.5 Case study: slippage sensing and reflex response

In Nakagawa-Silva et al. [11], they used an artificial hand (i-Limb prosthetic hand, Össur, USA) to evaluate the proposed slip detection and suppression framework in real-world manipulation task (Fig. 9.5). The experiment's aim was to assess the ability of the slippage sensor to detect and control the slip events when an object is subjected to dynamic events.

Two tasks were developed. The first task was for the sensing and control system to detect and adjust the finger when an object is dropped inside the bottle that the fingers are holding. The second task was for the system to respond to the increasing weight of water that is being poured into the bottle.

Both tasks made use of the i-Limb hand that was mounted as an end-effector of a robotic arm (UR10, Universal Robot, Denmark). The optical slip sensor was installed on the middle finger of the i-Limb hand. Additional tactile sensors were installed on the index finger and the thumb to monitor the contact forces. The results for the transparent (TR) bottle and a dark bottle (DK) will be reported in the next sections.

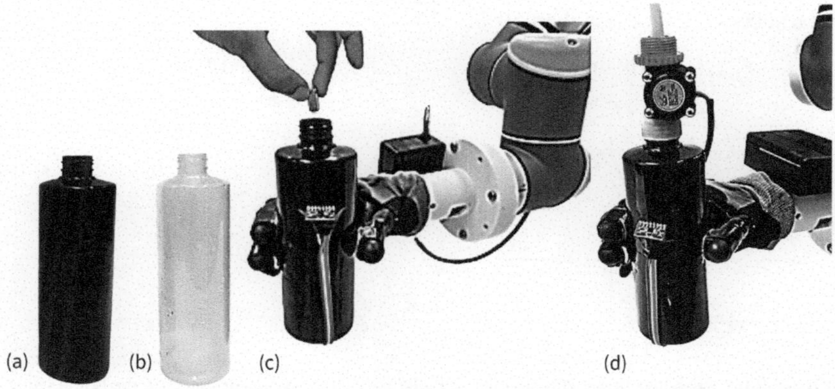

FIGURE 9.5 Bottles that were used in the experiment: (a) dark bottle and (b) transparent bottle. (c) In Experiment 1, a standardized object with a mass of 10 g was suddenly dropped into the bottle, triggering a fast slip event. (d) In Experiment 2, 500 ml of water was poured into the bottle to gradually increase the weight. A water pump was used with a constant flow rate (1 L/min). Adapted from Nakagawa et al. [11], with permission. ©2019 IEEE.

9.5.1 Slip detection when a mass is dropped

9.5.1.1 Methodology

The i-Limb hand was tasked with grasping a bottle using a tripod grasp configuration, engaging the thumb, index, and middle fingers. A standardized object weighing 10 g was dropped into the bottle from a height of 2 cm above its opening, simulating a sudden perturbation that would induce a rapid slip event. An Inertial Measurement Unit (IMU) was attached to the bottle to detect the onset of slip, and the accelerometer data was processed to identify slip events based on a two-step threshold detection method.

The first step involved determining a threshold based on the average and standard deviation of the initial 200 ms of the accelerometer signal. The bottle was stationary at this point. Samples exceeding this threshold were flagged as potential slip onsets. The second step involved evaluating these potential onsets. Only the first onset occurring within the time interval when the object was dropped was considered as the true slip onset.

The response time of the slip detection and suppression system was measured as the time elapsed between the slip onset and the first change in the i-Limb's finger position. This indicated the system's reaction to the slip event. The total control time was defined as the time elapsed from the initial response of the controller to the point where the slip was fully suppressed.

9.5.1.2 Results

For the sudden perturbation task, the DATH method achieved an average success rate of 80% in preventing complete slippage, while the fixed threshold method had a success rate of less than 50%.

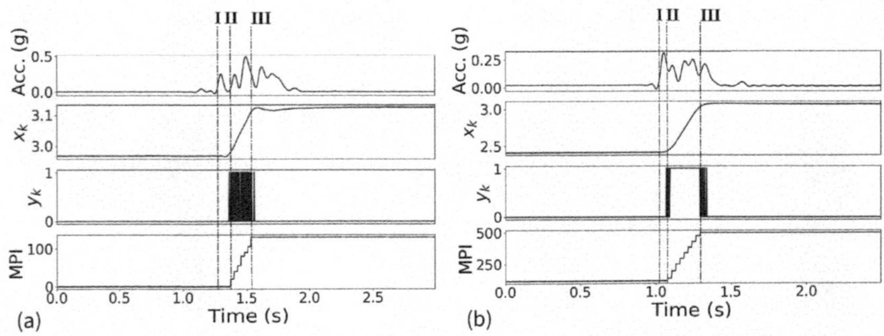

FIGURE 9.6 Response of the slip suppression when a 10 g mass was dropped on the bottle using the DATH for (a) the dark bottle and (b) the transparent bottle. The first row from the top shows the *y*-axis signal from the accelerometer. The second row shows the slip sensor's filtered output. The third row shows the events triggered by the slip of the bottle after the 10 g object was dropped. The last row shows the MPI controller's output. The dashed lines in (I) denote the system's response time as measured to be the elapsed time from the onset of the slip signal detected from the accelerometer signal until dashed lines in (II) which is the first response of the controller. The total time was measured as the time elapsed from the first response of the controller in (II) until the dashed line in (III) which is its final response when the slip was suppressed. Reproduced from Nakagawa et al. [11], with permission. ©2019 IEEE.

The response times were also measured (Fig. 9.6). The DATH method had average values of 50 ± 6.48 ms for the transparent bottle and 88.2 ± 24 ms for the dark bottle. These response times were comparable to human reaction times for grip force adjustments. In humans, the response time typically occurs within 60-80 ms after slip onset [3]. In contrast, the fixed threshold method exhibited slower response times.

The total time required for slip suppression was also significantly lower with the DATH method. The DATH method showed an average of 298.5 ± 77.41 ms for the transparent bottle compared to 661 ± 283 ms for the neuromimetic slip detection (NMSD) method described in [18].

9.5.2 Slip detection when a liquid is poured

9.5.2.1 Methodology

The i-Limb hand held the bottle using the same tripod grasp configuration. The weight of the bottle was gradually increased with a steady 1 L/min flow that filled the bottle in 500-ml increments. The gradual increase in weight created a continuous slip condition. A successful slip detection and control system should be able to respond to such conditions.

9.5.2.2 *Results*

The initial slip event typically occurred at around 200 ml of water while the second slip event was observed at approximately 400 ml level (Fig. 9.7). The Monotonic Proportional-Integral (MPI) controller successfully adjusted the grip force in response to these slip events. This allowed a stable grasp of the object without excessive grip force being applied.

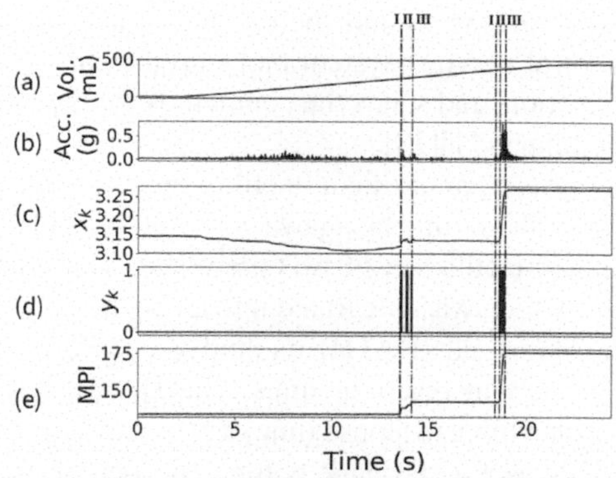

FIGURE 9.7 Response of the DATH slip suppression when a 500 ml volume of water was poured on the dark bottle. (a) The gradual increase of the water coming from the water flow sensor. (b) Accelerometer's output showing the *y*-axis signal from the accelerometer. (c) The slip sensor's filtered output. (d) The events triggered by the slip of the bottle. (e) The MPI controller's output. Similar to the previous results, the dashed lines in (I) denote the system's response time as measured to be the elapsed time from the onset of the slip signal detected from the accelerometer signal until dashed lines in (II) which is the first response of the controller. The total time was measured as the time elapsed from the first response of the controller in (II) until the dashed line in (III) which is its final response when the slip was suppressed. The gradual increase of the bottle's weight resulted in two instances of slippage. Reproduced from Nakagawa et al. [11], with permission. ©2019 IEEE.

The DATH method continued to outperform the fixed threshold approach. The total time required to suppress slip during the gradual weight increase was also significantly lower with the DATH method. The average time for the dark bottle was 460 ± 168 ms. The NMSD method recorded a slip suppression time of 749 ± 281 ms.

9.6 Summary

Understanding the forces acting on an object when it has been lifted requires us to consider two things: (a) the magnitude of the grasping force required to hold the object without causing damage or deformation, and (b) the magnitude of the force to counteract the object's weight to prevent it from slipping or falling.

The human tactile sensing system monitors the ratio of the grip force and load force to prevent an object from slipping. Learning the basic principles from this system allows us to design an artificial tactile sensing system that can sense slippage and control the grasping force on robotic or prosthetic fingers.

This chapter explored the mechanisms of object slippage sensing and the corresponding reflex responses. The chapter briefly described the human physiology of slippage sensing. The Meissner corpuscles (FA I) and Pacinian corpuscles (FA II) are responsible for detecting subtle changes in contact and enabling real-time adjustments in grip force to prevent slippage [3][4][2][5].

The two-phase pattern of slippage was also described; an initial phase of incipient slip events occurs at the edges of the contact area. These are detected by the human tactile sensing system before the object experiences full slip [9][6][7][8]. The central nervous system picks up the early incipient slippage, processes this information, and sends the commands for the fingers to modulate the grip force of the fingers [1][10].

Inspired by that system, an artificial tactile sensing system for slip sensing and control was described. In Nakagawa-Silva et al. [11], the design and evaluation of an optical tactile sensor was presented. The sensor was intended to act as a slip sensor which was integrated into an artificial hand. The proposed Dynamic Adaptive Threshold (DATH) method was able to achieve response times comparable to human reaction times. The DATH method significantly improved slip detection with a response time of 50 ± 6.48 ms for the transparent bottle and 88.2 ± 24 ms for the dark bottle. The neural mechanisms of the human hand can make the adjustments within 60-80 ms after the onset of slippage [3].

Knowing the fundamental mechanism of the slippage sensing of humans can provide us guidance on how to replicate a comparable system for artificial fingers. The free body diagram was useful for identifying all the forces experienced by the fingers and an object. The control system block diagram illustrated the closed-loop feedback coming from the slip sensor. The results from the optical sensor and the DATH algorithm demonstrated a slip sensing and reflex response technique to address one of the basic requirements of tactile sensing for robotic and prosthetic hands.

References

[1] R.S. Johansson, J.R. Flanagan, Coding and use of tactile signals from the fingertips in object manipulation tasks, Nature Reviews Neuroscience 10 (2009 Apr) 345–359, https://doi.org/10.1038/nrn2621.

[2] M.J. Adams, S.A. Johnson, P. Lefèvre, V. Lévesque, V. Hayward, T. André, J.L. Thonnard, Finger pad friction and its role in grip and touch, Journal of the Royal Society Interface 10 (2013) 20120467, https://doi.org/10.1098/rsif.2012.0467.

[3] R.S. Johansson, G. Westling, Roles of glabrous skin receptors and sensorimotor memory in automatic control of precision grip when lifting rougher or more slippery objects, Experimental Brain Research 56 (1984) 550–564, https://doi.org/10.1007/bf00237997.

[4] G. Cadoret, A.M. Smith, Friction, not texture, dictates grip forces used during object manipulation, Journal of Neurophysiology 75 (1996) 1963–1969, https://doi.org/10.1152/jn.1996.75.5.1963.

[5] L. Willemet, K. Kanzari, J. Monnoyer, I. Birznieks, M. Wiertlewski, Initial contact shapes the perception of friction, Proceedings of the National Academy of Sciences of the United States of America (2021) 118:e2109109118, https://doi.org/10.1073/pnas.2109109118.

[6] V. Levesque, V. Hayward, Experimental Evidence of Lateral Skin Strain During Tactile Exploration, in: Proceedings of EuroHaptics 2003, Dublin, Ireland, vol. 6, 2003 Jul, pp. 261–275.

[7] M. Tada, T. Kanade, An imaging system of incipient slip for modelling how human perceives slip of a fingertip, The 26th Annual International Conference of the IEEE Engineering in Medicine and Biology Society, vol. 3, IEMBS-04, IEEE, 2004, pp. 2045–2048, https://doi.org/10.1109/iembs.2004.1403601.

[8] B. Delhaye, P. Lefèvre, J.L. Thonnard, Dynamics of Fingertip Contact During the Onset of Tangential Slip, Journal of the Royal Society Interface 11 (2014) 20140698, https://doi.org/10.1098/rsif.2014.0698.

[9] A. Barrea, B. Delhaye, P. Lefèvre, J.L. Thonnard, Perception of partial slips under tangential loading of the fingertip, Scientific Reports 8 (2018) 7032, https://doi.org/10.1038/s41598-018-25226-w.

[10] B.P. Delhaye, M.R. Longo, P. Lefèvre, J.L. Thonnard, High-resolution imaging of skin deformation shows that afferents encode grip-relevant signals during object manipulation, eLife 10 (2021) e64679, https://doi.org/10.7554/eLife.64679.

[11] A. Nakagawa-Silva, N.V. Thakor, J.J. Cabibihan, A.B. Soares, A Bio-Inspired Slip Detection and Reflex-Like Suppression Method for Robotic Manipulators, IEEE Sensors Journal 19 (2019) 12443–12453, https://doi.org/10.1109/jsen.2019.2939506.

[12] R.S. Johansson, U. Landström, R. Lundström, Responses of mechanoreceptive afferent units in the glabrous skin of the human hand to sinusoidal skin displacements, Brain Research 244 (1982) 17–25, https://doi.org/10.1016/0006-8993(82)90899-X.

[13] A.M. Gordon, G. Westling, K.J. Cole, R.S. Johansson, Memory representations underlying motor commands used during manipulation of common and novel objects, Journal of Neurophysiology 69 (1993) 1789–1796, https://doi.org/10.1152/jn.1993.69.6.1789.

[14] A. Prach, J.J. Cabibihan, N.V. Thakor, D.S. Bernstein, Pareto-front analysis of a monotonic PI control law for slip suppression in a robotic manipulator, in: 2017 IEEE International Conference on Robotics and Biomimetics (ROBIO), IEEE, 2017, pp. 2728–2733, https://doi.org/10.1109/robio.2017.8324832.

[15] M. Al Janaideh, D. Bernstein, Adaptive control of mass-spring systems with unknown hysteretic contact friction, in: 2016 American Control Conference (ACC), IEEE, 2016, pp. 667–672, https://doi.org/10.1109/acc.2016.7524990.

[16] E.D. Engeberg, S.G. Meek, Adaptive Sliding Mode Control for Prosthetic Hands to Simultaneously Prevent Slip and Minimize Deformation of Grasped Objects, IEEE/ASME Transactions on Mechatronics 18 (2013) 376–385, https://doi.org/10.1109/TMECH.2011.2179061.

[17] A. Nakagawa-Silva, S.P. Reddy Sunkesula, A. Prach, J.J. Cabibihan, N.V. Thakor, A.B. Soares, Slip suppression in prosthetic hands using a reflective optical sensor and MPI controller, in: 2018 IEEE Biomedical Circuits and Systems Conference (BioCAS), IEEE, 2018, pp. 1–4, https://doi.org/10.1109/biocas.2018.8584711.

[18] L. Osborn, H. Nguyen, J. Betthauser, R. Kaliki, N. Thakor, Biologically inspired multi-layered synthetic skin for tactile feedback in prosthetic limbs, in: 2016 38th Annual International Conference of the IEEE Engineering in Medicine and Biology Society (EMBC), IEEE, 2016, pp. Aug:4622–5, https://doi.org/10.1109/embc.2016.7591757.

Chapter 10

Sensing of local surface primitives

10.1 Introduction

An object's physical features, such as local surface features, texture, stiffness, and temperature, can be best obtained by the sense of touch. When the hands touch objects, the tactile information from the force and pressure patterns are generated. These patterns can be used to obtain knowledge about an object that can be subsequently used for grasping and manipulation. The information about local surface primitives, such as the surface curvature of objects, can help us determine the orientation and configuration of the fingers for properly applying the grasping forces on the object [1].

The overall shape of objects can be determined globally or locally by minimizing their elements into curves or angles [2]. Global shape will be discussed in the next chapter. An object's global shape can be determined easily through the eyes, yet there are many situations in our daily lives that do not rely on vision. Daily tasks, such as buttoning a shirt, searching for coins in our pockets, shifting gears in the car, playing the guitar, or turning on the light switch in a dark room, do not need vision for us to complete the desired task.

Local surface primitives are basic geometric entities defined over the local neighborhoods of an object. The typical local surface primitives include curve surface patches, edges, or planar surface patches that originate from classical surface-based object classifications for three-dimensional object representation and recognition [3]. In this chapter, an object's local surface primitives, such as curvatures, edges, and flat surfaces are investigated from the lens of human and artificial tactile sensing. Among these, curvature is invariant to an object's translation or rotation. For humans, the minute differences in the curvatures of surfaces can be discriminable [4]. Discrimination among surface features would require us to understand curvatures better.

The human mechanoreceptors play a key role in the discrimination of a surface's fine features. During the exploration of a surface or an object's features, static contact by the fingers activates the mechanoreceptors through the deformation of the skin. LaMotte and Srinivasan [5] suggest that the static contact between the finger and an object determines the local curvature. When dynamic exploratory movements are made, the sliding or stroking movements give information on the texture of a surface and other fine details.

Humanizing Artificial Touch. https://doi.org/10.1016/B978-0-32-395519-5.00021-0

Sec. 10.2 presents an overview of how the human tactile sensing system discriminates the typical local surface geometry. Sec. 10.3 presents an artificial tactile sensing approach for discrimination of curved, edged, and flat surfaces. Sec. 10.4 gives a summary of the chapter.

10.2 Human tactile sensing of local surface primitives

How does the human hand differentiate the finest details of a surface feature or texture with touch alone? Differentiating a sharp edge, a bottle's curvature, or a flat plate requires the complex interplay between the biomechanics of the skin, the specialized mechanoreceptors, and the processing in the brain.

10.2.1 Skin biomechanics

Before a neural signal is generated, an object's physical shape or surface feature has to be touched through the skin. The skin converts the mechanical signal to a neural signal. It acts as a preprocessor for filtering mechanical signals for the mechanoreceptors to process.

The fingerprint ridges were commonly known to improve the friction between an object and the fingertip. These ridges also serve as mechanical amplification for the transduction of stresses and strains to the subsurface mechanoreceptors. Prevost et al. [6] showed that the fingerprints perform spectral selection and amplification of tactile information.

Phillips and Johnson [7] and Dandekar et al. [8] suggested that the quantity to stimulate the mechanoreceptors is the strain energy density (SED), which is a measure of the mechanical energy stored in the tissue that was deformed. The object's geometry influences the pattern of that subsurface energy. A curved object with a small radius of curvature creates a smaller contact area (see Fig. 10.1). This results in higher contact pressure. The high contact pressure results in a concentration of SED at the contact point. On the contrary, when a flat object is in contact with the skin, the same force is now distributed in a larger area. The contact is more uniform, which results in less SED.

10.2.2 Mechanoreceptors

The slowly adapting (SA) receptors were said to contribute to the discrimination of shapes or surface features. The SA I mechanoreceptors provide the spatial information necessary to detect fine details like edges and curvature [10]. The SA II mechanoreceptors provide information related to skin stretch, overall shape, force magnitude, and the posture of the hand [11].

In the tactile curvature experiments of Goodwin et al. [12], they showed that the SA I afferents are the main mechanoreceptors that encode an object's static shape or geometric feature. They explained that the nervous system makes use of population coding. The brain analyzes the spatial pattern throughout the entire

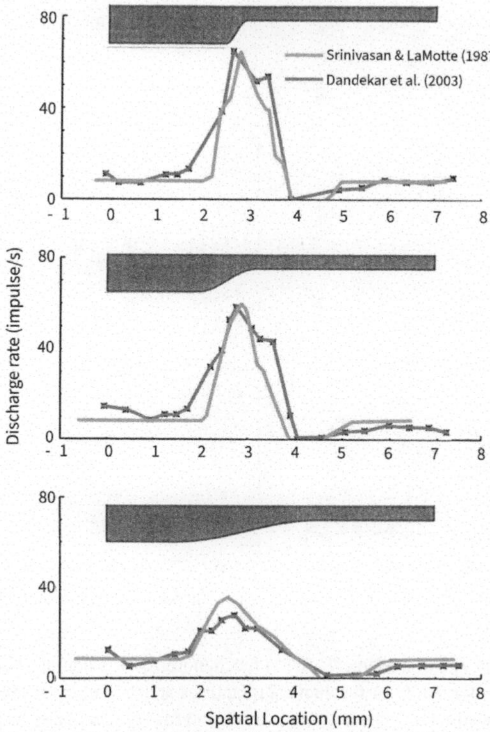

FIGURE 10.1 Indentation by stepped shapes with curved profiles. Comparison of the Slowly Adapting Type 1 (SA I) mechanoreceptor's neural response by Srinivasan and LaMotte [9] and the modeling results with strain energy density by Dandekar et al. [8]. Adapted from Dandekar et al. [8], with permission. ©1999 ASME. All rights reserved.

population of the afferents that were activated with the stimulus. The signals are then integrated from the different receptors.

In the studies of Phillips and Johnson [7], they indented the fingerpad with gratings. Neurophysiological data on the response of SA I mechanoreceptors were obtained. Fig. 10.2 shows the discharge rate of SA I mechanoreceptors for a 3 mm wide rectangular bar. The peaks of the discharge rate correspond to a width of less than 4 mm. Dandekar et al. [8] modeled a fingertip with finite element analysis. After indenting the finger with the 3 mm-wide bar, they plotted the SED, maximum compressive strain, mean normal stress, and vertical compressive stress. Among these, the SED was able to predict the neural response of the SA I mechanoreceptor.

10.2.3 Discrimination among planar, edged, and curved surfaces

In the discrimination of a flat surface vs. an edge, the deformation of the skin helps in the preprocessing of the signals. Phillips and Johnson [7] showed that an

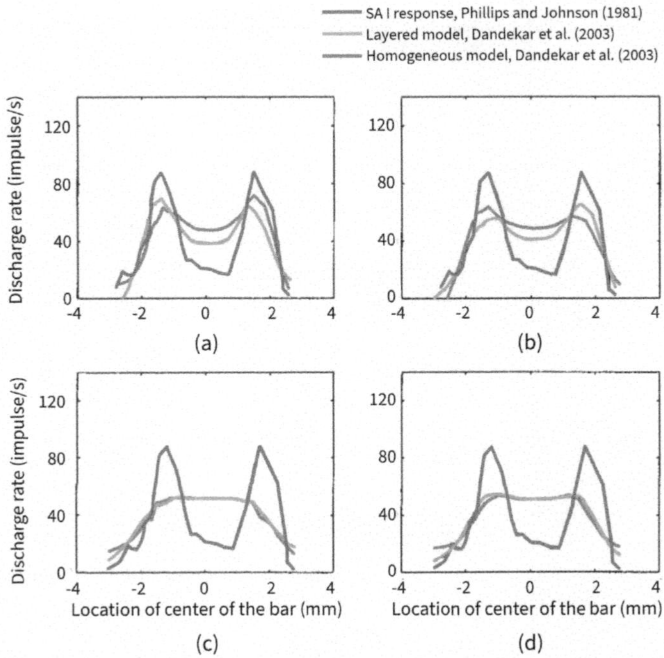

FIGURE 10.2 Slowly Adapting Type 1 (SA I) mechanoreceptor's neural response to indentation by 3 mm-wide rectangular bar. Four mechanical quantities are shown: (a) strain energy density, (b) maximum compressive strain, (c) mean normal stress, and (d) vertical compressive stress. The plots were compared for a layered finite element model with a layered skin and a homogeneous skin and against the model of Phillips and Johnson [7]. The *x*-axis is the location of the center of the bar. Adapted from Dandekar et al. [8], with permission. ©1999 ASME. All rights reserved.

edge enhancement occurs due to the concentration of stresses and strains. They showed that the response profile of SA I to a narrow bar or edge is enhanced at the discontinuities.

Wheat and Goodwin [13] further demonstrated that SA I responses show asymmetry at the edges. The neural data from their experiments showed that the information on a specific surface feature of an edge is encoded by the distributed spatial signals within the afferent population.

In the discrimination of a flat surface vs. a curved one, the effect of the contact area becomes important to consider. A flat surface, which has a curvature of infinite radius, has a distribution of force in a wider area of the skin. This results in lower peak pressure and a population response profile that is flat. On the contrary, a curved surface, which has a decreasing radius, has a smaller contact area and a higher local stress.

Goodwin et al. [12] demonstrated that the increase of the curvature on the fingerpads increases the firing rates of the SA I mechanoreceptors. They also showed that the response profile of the population keeps a stable shape but scales in intensity. The neural image of a curved surface appears as a hill of activity where the height (i.e., frequency) and the width encode the curvature and contact force.

They further add that the central nervous system (CNS) can discriminate a curve from a flat surface by monitoring the peak firing rates and the spatial characteristics of the active SA I population. In short, a lower and broad response indicates a flat surface while a high-peaked, spatially confined response indicates a curved object.

10.3 Artificial tactile sensing of local surface primitives: case study

In Salehi et al. 2011 [14], we described a method that aimed to address the limitations of current artificial hands in comparing local surface primitives, specifically surface curvature, which is an important input for dexterous manipulation.

The primary design objective was to enhance the curvature detection capabilities of embedded tactile sensors by introducing a novel ridged skin cover mimicking human fingerprints. This design takes inspiration from the human fingerprints in enhancing tactile sensitivity and information processing.

A secondary objective was to develop a machine learning-based method for classifying surface curvature from tactile sensor data, overcoming limitations of previous approaches that often relied on vision-based tactile image processing, which has drawbacks in terms of high memory usage and slower processing speeds.

10.3.1 Methodology

A three-step approach was used to investigate the discrimination of local surface primitives from the outer layer of flat skin or ridged skin (Fig. 10.3a-10.3b). The process involved (1) indentation experiments, (2) elimination of noise and repeatability evaluation, and (3) machine learning and analysis.

For the indentation experiments, a tactile sensor array was used (TactoPad, TactoLogic Inc., Hungary; Fig. 10.3c). The tactile array was comprised of four three-dimensional taxels on a 4 mm $\times$ 4 mm active area, each measuring normal and tangential contact forces with sensitivities of 5 mN/bit and 2 mN/bit respectively. Two silicone rubber (GLS 40, Prochima, Italy) skin covers were employed: a flat cover and a ridged cover, both designed to accommodate the sensor's rubber bumps (Fig. 10.3d).

Silicone rubber was chosen for its biocompatibility and compliance. Three steel objects representing distinct surface curvatures were used as stimuli: flat (15 $\times$ 5 mm^2; Fig. 10.3e), edged (5 mm length, 0.1 mm edge width; Fig. 10.3f), and curved (0.4 mm^{-1} curvature, 2.5 mm radius; Fig. 10.3g).

The indenters were directed toward the center of the tactile array so that consistent contact can be made (Fig. 10.4). Forty indentations were set for each

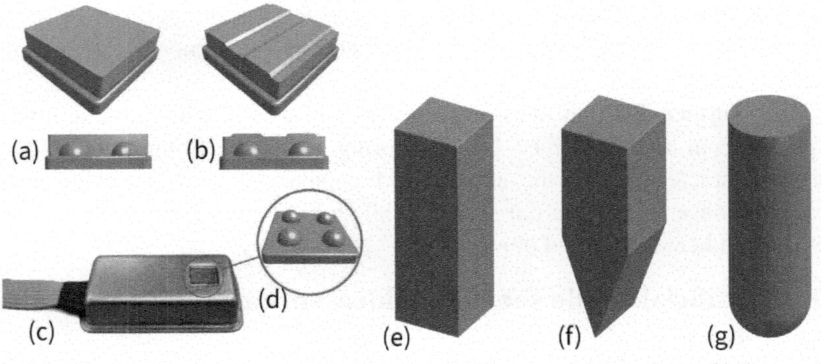

FIGURE 10.3 Tactile sensor array with two types of skin covers and three types of indenters. (a) Artificial skin with a flat cover as a baseline to verify the effect of ridges. (b) The ridged skin cover on the tactile sensor array. (c) The tactile sensor consisting of 2 × 2 array on a 4 mm × 4 mm area. Each sensor measures the contact force in normal and tangential directions. (d) The magnified version of the tactile array. (e) Experimental object with flat surface of 5 mm × 5 mm. (f) Sharp edge with the length of 5 mm and thickness of 0.1 mm. (g) Convex curved surface with a curvature of 0.4 mm^{-1} (radius of 2.5 mm).

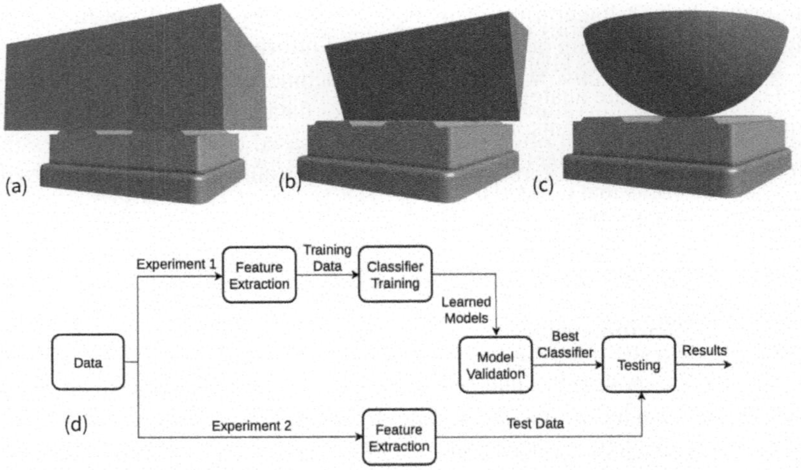

FIGURE 10.4 Illustration of the indentation process on the tactile array with the ridged skin cover. The sensor with ridged skin cover with (a) flat surface, (b) sharp wedge, and (c) curved surface. (d) The learning procedure with the top path showing the classifier training using the collected data from Experiment 1 and the bottom path displaying the testing of the trained model using collected data from Experiment 2. Adapted from Salehi et al. [14]. ©2011 Salehi et al. under the terms of the Creative Commons Attribution 4.0 International License.

surface primitive (i.e., planar, edged, or curved) and for each skin cover (i.e., flat or ridged). The loading condition was a 0.5 N/s indentation (Phase A), steady state load of 2 N for 2 s (Phase B), and unloading with a rate of 0.5 N/s (Phase C). The signal processing step involved a low-pass filter (i.e., Finite Impulse Response) with a 1 Hz cut-off frequency. This addressed the high sensitivity and potential noise in the force data from the tactile sensors.

The filter processed the entire data window for each phase of the indentation process. The repeatability of the measurements was assessed using pairwise Pearson cross-correlation of the time series from the four sensors, similar to the work in [15]. The Pearson cross-correlation coefficient was defined as:

$$r_{xy} = \frac{\sum_{i=1}^{n}(x_i - \overline{x})(y_i - \overline{y})}{\sqrt{\sum_{i=1}^{n}(x_i - \overline{x})^2}\sqrt{\sum_{i=1}^{n}(y_i - \overline{y})^2}} \tag{10.1}$$

where $\overline{x}$ and $\overline{y}$ are the average values of the x and y pair, respectively. The size of the measurement time series is denoted as n.

The last step, the learning and analysis step, involved feature extraction, training, validation, and testing. Data were extracted from the force signals of all four sensors. These data involved statistical features, such as maximum value, mean, standard deviation, skewness, and kurtosis. These were then normalized using Min-Max normalization [16] before being used to train and validate machine learning algorithms:

$$x_{new} = \frac{x_{old} - x_{min}}{x_{max} - x_{min}}(max_{desired} - min_{desired}) + min_{desired} \tag{10.2}$$

where $max_{desired}$ is the desired maximum value (i.e., 1) and $min_{desired}$ is the desired minimum value (i.e., 0).

We used Artificial Neural Networks (ANN), Naïve Bayes (NB), and Support Vector Machines (SVM) to evaluate the data. The different learning methods were compared. The accuracy, calculated as the ratio of correctly classified samples to the total number of samples, served as the performance metric. This was defined as:

$$Accuracy = \frac{N_c}{N_t} \tag{10.3}$$

where the number of correctly predicted samples in the validation phase is denoted as N_c and the total number of samples is denoted as N_t.

10.3.2 Results

The indentation experiments generated force profiles using a 2 × 2 tactile sensor array with both flat and ridged silicone rubber skin covers. Three distinct objects, curved (Fig. 10.5a), edged (Fig. 10.5b), and planar (Fig. 10.5c) were used as stimuli.

The noise elimination and repeatability assessment involved applying a designed FIR low-pass filter with a cut-off frequency of approximately 1 Hz to reduce noise in the force signals. Fig. 10.6a and 10.6b show examples of the filtered force signals for both flat and ridged skin covers indented with a flat surface, highlighting the effectiveness of the filter in smoothing the raw data. The repeatability of the measurements was assessed using pairwise

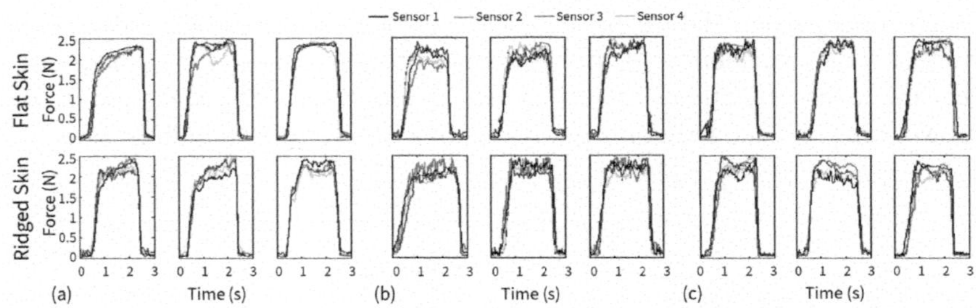

FIGURE 10.5 Sample force data results for flat and ridged skin covers when indented with the following surface primitives: indenters consisting of (a) curved, (b) edged, and (c) planar. The results come from the resultant forces from the four sensors of the tactile sensor array. Adapted from Salehi et al. [14]. ©2011 Salehi et al. under the terms of the Creative Commons Attribution 4.0 International License.

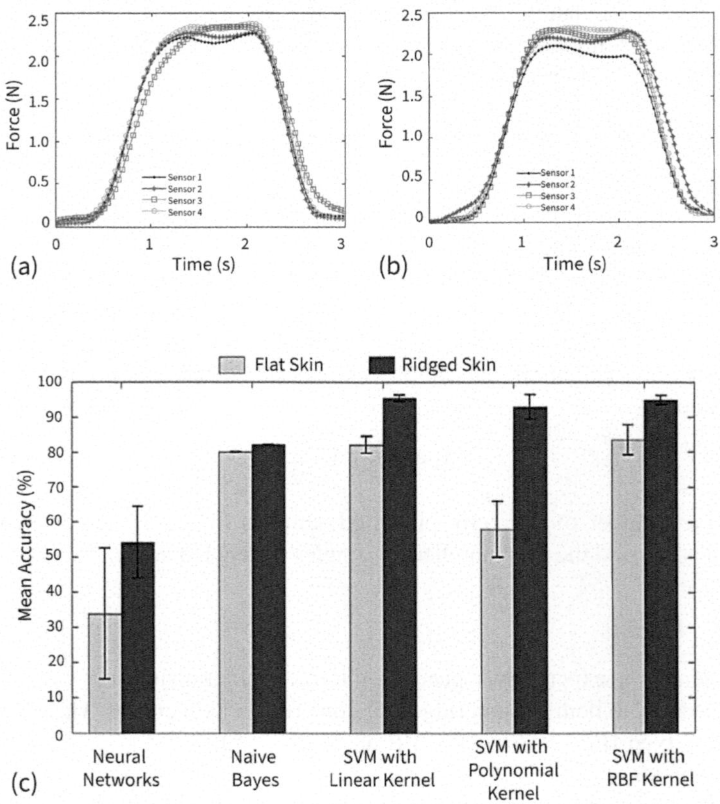

FIGURE 10.6 Resultant forces from the 2 × 2 sensor array for (a) the flat skin cover and flat surface indenter and, (b) the ridged skin cover and flat surface indenter. (c) The comparison of various machine learning algorithms with the light gray bars representing the mean value of algorithms' accuracy rates with flat skin cover and the dark gray showing the mean accuracy rate for the ridged one. Reproduced from Salehi et al. [14]. ©2011 Salehi et al. under the terms of the Creative Commons Attribution 4.0 International License.

TARGET / OUTPUT	Curved	Edged	Flat	SUM
Curved	39 32.50%	0 0.00%	1 0.83%	40 97.50% 2.50%
Edged	1 0.83%	39 32.50%	0 0.00%	40 97.50% 2.50%
Flat	1 0.83%	0 0.00%	39 32.50%	40 97.50% 2.50%
SUM	41 95.12% 4.88%	39 100.00% 0.00%	40 97.50% 2.50%	117 / 120 97.50% 2.50%

FIGURE 10.7 Confusion matrix for the training set (n = 120) of the three-class contact-surface primitive classifier (curved, edged, and flat). Columns are ground truth and rows are predicted labels; diagonal cells indicate correct classifications. Overall accuracy is 97.5% (117/120); per-class recall: curved 95.1%, edged 100%, and flat 97.5%. Reproduced from Salehi et al. [14]. ©2011 Salehi et al. under the terms of the Creative Commons Attribution 4.0 International License.

Pearson cross-correlation coefficients (Eq. (10.1)). Calculations for the Pearson correlation coefficient showed high average values that exceed 99%. This confirms a high degree of consistency and reliability in the data collection.

Three machine learning algorithms, ANN, NB, and SVM, were employed, with SVM demonstrating superior performance overall. The mean accuracy rates for the various machine learning algorithms used are shown in Fig. 10.6c. The light gray bars show the results for the flat skin cover, while the dark gray bars show the results for the ridged skin. The ridged skin performs better than the flat skin cover. SVM exhibited better performance compared to the other algorithms. With the ridged skin and the SVM with RBF kernel showing the best performance, only the test data for the ridged skin cover was evaluated for the test phase.

Discrimination was achieved with an accuracy rate of 97.5% for the surface curvatures of the test dataset. The confusion matrix in Fig. 10.7 shows the number of samples that were correctly detected for each surface primitive. From the 120 samples collected with ridged cover for testing, the first 40 samples are for curved, the second 40 are for edged, and the third set of 40 samples are for the flat surface.

10.4 Summary

Understanding the basic sequence and process of how the human tactile sensing system detects and perceives surface features guided us with a solution for an artificial tactile sensing system. By including a ridged skin cover over a 2×2 array of sensors that detect forces in the *x*-, *y*-, and *z*-axes, we were able to see that the signals that have been transmitted to the embedded sensors have been mechanically preprocessed.

The mechanoreceptors in the fingertip skin have the marked ability to encode the static surface features [17][5][18]. SA I mechanoreceptors can detect curvature differences of up to 0.1 mm [19]. The fingertip has a high concentration of these mechanoreceptors of around (70–100 per cm^2) [20]. The human tactile perception of surface features is further enhanced by the skin's mechanical filtering properties [21]. Integrating all of these gives cues to allow any modulation that is needed for a successful grasping and manipulation of the human hand [1][12].

This chapter presented a study on a 2×2 tactile sensor array, which compared the results of flat and ridged silicone skin covers to evaluate their use for discriminating surface primitives [14]. The methodology included the following steps. First, experiments were conducted that indented a planar surface, wedged indenter, and curved indenter on the sensor with flat or ridged skin. Second, noise was eliminated and the repeatability of the results were assessed. Lastly, the machine learning and analysis procedures were conducted. Using the tactile sensor data, a model was trained and then the learned model was applied to new sensory data to detect the local surface geometry.

The machine learning methods used were Neural Networks, Naïve Bayes, Support Vector Machines (SVMs) with a linear kernel, polynomial kernel, and radial basis function (RBF) kernel. The SVM was shown to be the most efficient algorithm that was able to classify the force profiles from the sensors' responses. More specifically, the SVM with RBF kernel was found to achieve the highest accuracy of 97.5% success rate for the ridge skin results.

It has been known that the minute details of the human fingerprint ridges have the ability to provide spectral selection of force signals [6]. This knowledge gave us an insight into a mechanical intelligence for an artificial tactile sensing, whereby we can add a ridged silicone skin above an array of triaxial force sensors so that the sensors can discriminate local surface features, such as a plane, edged, or curved surfaces from one another.

References

[1] P. Jenmalm, S. Dahlstedt, R.S. Johansson, Visual and tactile information about object-curvature control fingertip forces and grasp kinematics in human dexterous manipulation, Journal of Neurophysiology 84 (2000) 2984–2997, https://doi.org/10.1152/jn.2000.84.6.2984.

[2] E. Gentaz, Y. Hatwell, Haptic Processing of Spatial and Material Object Properties. Touching for Knowing, John Benjamins Publishing Company, 2003, pp. 123–159, https://doi.org/10.1075/aicr.53.12gen, Chapter 8.

[3] O. Faugeras, M. Hebert, The representation, recognition, and locating of 3-D objects, The International Journal of Robotics Research 5 (1986) 27–52, https://doi.org/10.1177/027836498600500302.

[4] A.M.L. Kappers, J.J. Koenderink, I. Lichtenegger, Haptic identification of curved surfaces, Perception & Psychophysics 56 (1994) 53–61, https://doi.org/10.3758/bf03211690.

[5] R.H. LaMotte, M.A. Srinivasan, Tactile discrimination of shape: responses of rapidly adapting mechanoreceptive afferents to a step stroked across the monkey fingerpad, Journal of Neuroscience 7 (1987) 1672–1681, https://doi.org/10.1523/JNEUROSCI.07-06-01672.1987.

[6] A. Prevost, J. Scheibert, G. Debrégeas, Effect of fingerprints orientation on skin vibrations during tactile exploration of textured surfaces, Communicative & Integrative Biology 2 (2009) 422–424, https://doi.org/10.4161/cib.2.5.9052.

[7] J.R. Phillips, K.O. Johnson, Tactile spatial resolution. II. Neural representation of bars, edges, and gratings in monkey primary afferents, Journal of Neurophysiology 46 (1981 Dec) 1192–1203, https://doi.org/10.1152/jn.1981.46.6.1192.

[8] K. Dandekar, B.I. Raju, M.A. Srinivasan, 3-D finite-element models of human and monkey fingertips to investigate the mechanics of tactile sense, Journal of Biomechanical Engineering 125 (2003) 682–691, https://doi.org/10.1115/1.1613673.

[9] M.A. Srinivasan, R.H. LaMotte, Tactile discrimination of shape: responses of slowly and rapidly adapting mechanoreceptive afferents to a step indented into the monkey fingerpad, Journal of Neuroscience 7 (1987) 1682–1697, https://doi.org/10.1523/JNEUROSCI.07-06-01682.1987.

[10] K.O. Johnson, The roles and functions of cutaneous mechanoreceptors, Current Opinion in Neurobiology 11 (2001) 455–461, https://doi.org/10.1016/s0959-4388(00)00234-8.

[11] P. Jenmalm, I. Birznieks, A.W. Goodwin, R.S. Johansson, Influence of object shape on responses of human tactile afferents under conditions characteristic of manipulation, European Journal of Neuroscience 18 (2003) 164–176, https://doi.org/10.1046/j.1460-9568.2003.02721.x.

[12] A.W. Goodwin, V.G. Macefield, J.W. Bisley, Encoding of object curvature by tactile afferents from human fingers, Journal of Neurophysiology 78 (1997) 2881–2888, https://doi.org/10.1152/jn.1997.78.6.2881.

[13] H.E. Wheat, A.W. Goodwin, Tactile discrimination of edge shape: limits on spatial resolution imposed by parameters of the peripheral neural population, Journal of Neuroscience 21 (2001) 7751–7763, https://doi.org/10.1523/JNEUROSCI.21-19-07751.2001.

[14] S. Salehi, J.J. Cabibihan, S.S. Ge, Artificial skin ridges enhance local tactile shape discrimination, Sensors 11 (2011) 8626–8642, https://doi.org/10.3390/s110908626.

[15] C.M. Oddo, L. Beccai, M. Felder, F. Giovacchini, M.C. Carrozza, Artificial roughness encoding with a bio-inspired MEMS-based tactile sensor array, Sensors 9 (2009) 3161–3183, https://doi.org/10.3390/s90503161.

[16] A. Jain, K. Nandakumar, A. Ross, Score normalization in multimodal biometric systems, Pattern Recognition 38 (2005 Dec) 2270–2285, https://doi.org/10.1016/j.patcog.2005.01.012.

[17] A.W. Goodwin, K.T. John, A.H. Marceglia, Tactile discrimination of curvature by humans using only cutaneous information from the fingerpads, Experimental Brain Research 86 (1991) 663–672, https://doi.org/10.1007/BF00230540.

[18] H.E. Wheat, A.W. Goodwin, A.S. Browning, Tactile resolution: peripheral neural mechanisms underlying the human capacity to determine positions of objects contacting the fingerpad, Journal of Neuroscience 15 (1995) 5582–5595, https://doi.org/10.1523/JNEUROSCI.15-08-05582.1995.

[19] I.E. Gordon, V. Morison, The haptic perception of curvature, Perception & Psychophysics 31 (1982) 446–450, https://doi.org/10.3758/BF03204854.

[20] R. Johansson, A. Vallbo, Tactile sensibility in the human hand: relative and absolute densities of four types of mechanoreceptive units in glabrous skin, Journal of Physiology 286 (1979) 283–300, https://doi.org/10.1113/jphysiol.1979.sp012619.

[21] M. Shimojo, Mechanical filtering effect of elastic cover for tactile sensor, IEEE Transactions on Robotics and Automation 13 (1997) 128–132, https://doi.org/10.1109/70.554353.

Chapter 11

Sensing of global shape primitives

11.1 Introduction

To achieve a successful grasping, manipulation, and recognition of objects, an object's overall form or global shape has to be perceived. In the previous chapter, the discrimination of local shape primitives was presented. These primitives can be determined by touching the surfaces locally through the fingertips. For larger object geometries, both cutaneous information from the skin mechanoreceptors and proprioceptive information from the muscles, tendons, and joints need to be integrated for the brain to construct a neural image of an object's shape, even without visual cues [1][2].

Global shape primitives, formally defined as compact volumetric models by Solina and Bajcsy [3], are parameterized geometric representations used to describe an object's gross structure and spatial extent. The typical global shape primitives include cylinders, spheres, ellipsoids, and parallelepipeds (i.e., boxes). Solina and Bajcsy [3] extended the classification of those shapes using deformable superquadrics. These are parametric models that are capable of global deformations, like tapering and bending. All these primitives are typically used in computer vision and robotics for object recognition, grasp planning, and pose estimation.

This chapter briefly describes the mechanisms of the human tactile sensing system to perceive global shapes. The chapter explores how humans combine cutaneous and proprioceptive information to comprehend an object's overall geometry. The exploratory procedures for acquiring shape information and the existence of the so-called "What" and "Where" of haptic processing are described.

Analogous to human experiments where digital anesthesia was used in selectively eliminating cutaneous feedback to investigate the role of proprioception [4], this chapter presents a case study where only proprioception is investigated. A sensorized haptic glove was used to test whether proprioception alone can discriminate among various shape primitives, such as a box, sphere, cone, or a cylinder.

Sec. 11.2 presents an overview of how the human tactile sensing system discriminates the overall shape of an object. Next, Sec. 11.3 presents a case study that demonstrated that discrimination among object shapes can be

Humanizing Artificial Touch. https://doi.org/10.1016/B978-0-32-395519-5.00022-2

achieved with proprioception. Finally, Sec. 11.4 concludes this chapter with a summary of the outcomes.

11.2 Human tactile sensing of global shape primitives

11.2.1 Exteroceptive sensing

Haptic perception of shape starts at the interface between the skin and the object. Exteroceptive or cutaneous sensing refers to the information coming from the mechanoreceptors at the glabrous part of the fingertip skin, without the involvement of the joints. As presented in Chap. 10, an object's local shape is best encoded by the Slowly Adapting type I (SA I) afferents for static contact. Goodwin et al. [5] showed that the firing rate of SA I afferents at the center of contact increases monotonically with the magnitude of local surface curvature. Consequently, this provides a robust signal of an object's shape at that area of contact. For example, Goodwin et al. [5] reported that a highly curved object results in a high and narrow response profile from the population of the SA I afferents. On the other hand, a flatter object results in a lower and wider profile.

The temporal structure of the neural signals is critical for the transfer of information of the local surfaces to the tactile afferents beneath the fingertip skin surface. Johansson and Birznieks [6] showed that the precise timing of action potentials in tactile afferents contains information about surface features, such as curved or flat surfaces, that can be obtained from the spike counts alone. The Fast Adapting type I (FA I) afferents are specialized mechanoreceptors for dynamic events, such as initial contact and return of objects, vibrations, and textures. The FA I afferents are effective for detecting discontinuities like edges and corners [7].

11.2.2 Proprioceptive sensing

Proprioception or kinesthesis refers to the sensing of limb position and movement information coming from the receptors in the muscle spindles, Golgi tendon organs, and the joint receptors (Fig. 11.1). When a hand grasps an object, the proprioceptors provide continuous information on joint angles and muscle states [1]. This information is integrated by the nervous system to derive the posture of the hand. The object's overall size and shape correspond to the postural information. For objects larger than the fingertip, the hand applies enclosure and the exploratory movement strategies to sense and perceive the global shape of an object.

Another mechanism, termed dynamic touch [9], makes use of proprioception to perceive an object's shape by actively moving the object. The mechanism relies on an object's moment of inertia tensor, which is the property that describes an object's resistance to rotational acceleration as a function of its mass distribution and geometry. In the experiments conducted by Burton et al. [10], they tested whether blindfolded participants could discriminate shapes

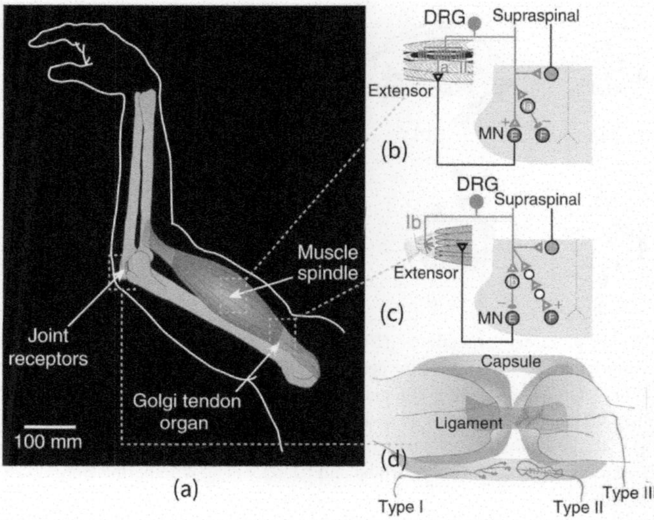

FIGURE 11.1 Proprioceptive sensors. (a) Locations of the muscle spindle, Golgi tendon organ, and joint receptors. (b) Illustration of the muscle spindle afferent spinal connectivity. (c) Illustration of the Golgi tendon organ. (d) Illustration of the receptor types in the joint capsule and ligament. Adapted from Tuthill and Azim [8], with permission.

from one another. The participants were presented with a pair of wooden objects (i.e., one in each hand). The mass of the objects was controlled to be as equal as possible. The object had a handle on it so that the participant's hands could not directly touch the actual shape. The objects consisted of a hemisphere, a cylinder, a rectangular block, a cone, and a pyramid. The participant was asked to compare whether the objects were the same or different. The objects could be moved in whichever motion the participants preferred. Results showed that a cylinder vs. a cone can be discriminated with a low error rate (i.e., 1/40 trials) while a cylinder vs. a box had a high rate of confusion (i.e., 16/40 trials).

11.2.3 Exploratory procedures

Exploratory procedures (EP) are specific, goal-directed motor strategies for one to acquire shape information [11]. Each EP is selected to link a perceptual goal (e.g., determine an object's shape) to a motor program that will maximize the acquisition of the necessary sensor information. Fig. 11.2 illustrates the EPs and the information that can be obtained from the active movements of the hands or fingers.

11.2.4 "What" and "where" in haptic processing

Neuroimaging studies by Reed et al. [12] confirmed the existence of two functionally and anatomically dissociated processing streams for the "What"

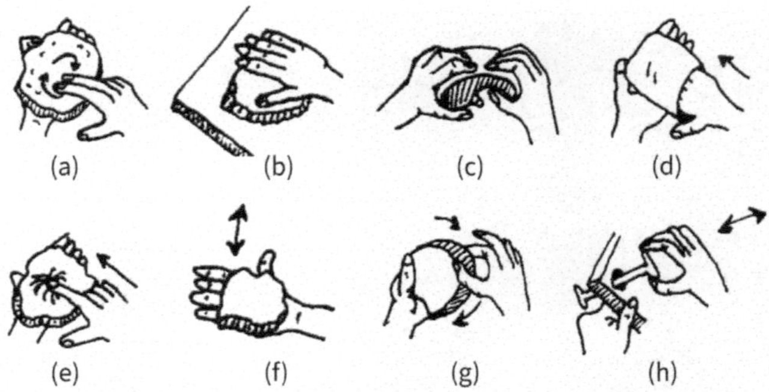

FIGURE 11.2 Exploratory procedures (EP) showing the typical patterns of movements in collecting sensory information as inputs for perceiving an object's shape, features, or purpose. These EPs are: (a) lateral motion for texture, (b) static contact for temperature, (c) enclosure for volume, (d) function test for specific function of an object, (e) pressure for hardness, (f) unsupported holding for weight, (g) contour following for volume and an object's exact shape, and (h) part motion test for an object's motion. Adapted from Lederman and Klatzky [11], with permission. ©1987 Elsevier. All rights reserved.

and "Where" of processing haptic information (Fig. 11.3). The "What" attributes include the object's shape, size, texture, and hardness. The "Where" attributes include the location of an object relative to the body. The "Where" pathway enables the reaching and grasping of an object in the environment.

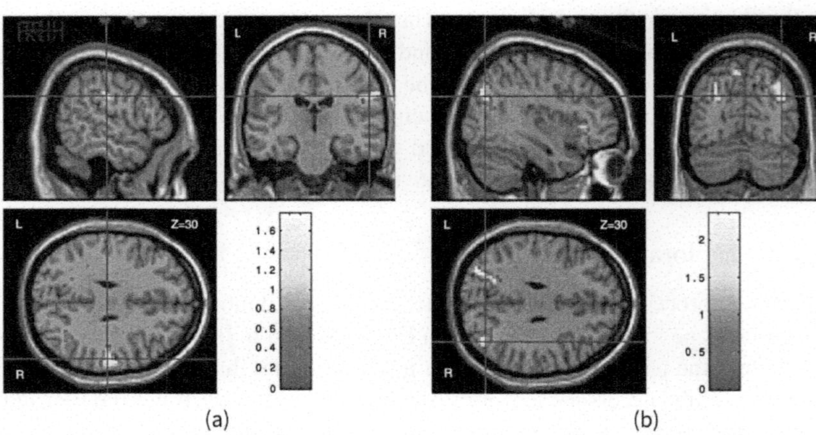

FIGURE 11.3 The functional magnetic resonance imaging (fMRI) to show the neural pathway for the "What" and "Where" processing in touch. (a) The "What" pathway to illustrate the Tactile Object Recognition (TOR) vs Movement (MOVT). The brain regions show the preferentially active locations during object identification. (b) The "Where" pathway to illustrate the Tactile Object Localization (LOC) vs Movement (MOVT). The regions showing the active regions during spatial processing. (L = Left, R = Right). The color bars indicate the statistical significance of the activation. Reproduced from Reed et al. [12], with permission. ©2005 Elsevier. All rights reserved.

The functional magnetic resonance imaging (fMRI) results in [12] showed that object identification (i.e., "What") selectively engages a ventrally-directed pathway, which includes the lateral occipital cortex. Tactile object localization tasks (i.e., "Where") activate a dorsally-directed pathway in the superior and inferior parietal lobules.

11.3 Artificial tactile sensing of global shape primitives: case study

In Anand et al. (2013) [13], we described a method to discriminate various object shapes using a sensorized glove to perform sensing similar to proprioceptive sensing in humans. The glove collects finger joint angles, which were then processed with machine learning techniques to discriminate a few objects. The primary design objective was to explore the feasibility of using solely kinesthetic information. In particular, this case study investigated the finger joint angles for the discrimination of primitive shapes. This contrasts sharply with existing methods which often rely on visual or combined visual-tactile data.

We sought to develop a model capable of recognizing various shapes based solely on the sensory input from a sensorized glove to investigate the role of artificial proprioceptors. The other objective was to determine which joints are providing significant results in differentiating primitive shapes. Each joint was analyzed in the overall classification of the accuracy.

11.3.1 Methodology

This case study aimed to evaluate the feasibility of using proprioceptive information through the flexion of the joint angles to discriminate various shape primitives. The methodology required two steps: data acquisition and model development.

11.3.1.1 Proprioceptive sensors

A sensorized glove (CyberGlove II, CyberGlove Systems, USA) was used for the data acquisition to replicate the proprioceptive sensors in the human hands. This glove is a wearable motion capture glove with 22 flexion sensors located at the joints. The movements that can be captured include joint flexion/extension, abduction/adduction, and wrist movements with an accuracy of 1°.

11.3.1.2 Model development and training

Two models were developed, which employed objects of varying shapes along with the corresponding grasping poses as needed to grasp those objects. Model 1 made use of four sets of shapes composed of a box, a cylinder, a flat surface, and a general grasp. These were chosen so that the dataset would have a balanced representation of the fundamental shape categories. Having a simple baseline allows the initial model training and evaluation to be simpler before modeling more complex objects. Multiple instances of grasps were obtained for each shape category.

Model 2 made use of more complex shapes for the discrimination task. This included: a ball, a box, a cone, a cylinder, a flat surface, a general grasp, and a ring. The ring-shaped object was further differentiated into three grasp patterns: a grip from the top, a bowling grip, and a general grasp. The use of multiple grasps for a single shape was able to increase the difficulty of the discrimination task. These had the benefit of using the discrimination's ability to handle ambiguities arising from various interaction scenarios with an object.

Similar to Model 1, the participants grasped the objects freely. The sensorized glove collected the joint angles. The dataset was much larger for Model 2 due to the increase in the number of objects and the additional variations in grasp poses.

The training of the models involved the classification of shapes from the acquired data which consisted of the finger joint angles. A Support Vector Machine (SVM) algorithm with radial basis function (RBF) kernel was used. SVM is a supervised learning model used for classification that finds an optimal hyperplane. This hyperplane maximally separates different shape classes in a high-dimensional feature space. The finger joint angles were mapped by the RBF kernel into a higher-dimensional space where a linear separation might be possible even if the original data is not linearly separable.

11.3.1.3 Algorithm

The optimization problem for a soft-margin SVM [14] is given by:

$$\begin{aligned} &\min_{w,\,b,\,\xi} \quad \tfrac{1}{2} w^T w + C \sum_{i=1}^{l} \xi_i \\ &\text{subject to} \quad y_i(w^T \varphi(x_i) + b) \geq 1 - \xi_i, \xi_i \geq 0, \end{aligned} \tag{11.1}$$

where $C > 0$ is the penalty factor of the error term. The training vectors x_i are mapped into a higher dimensional space by the function φ. Different φ kernels like linear, polynomial, or radial basis functions, may be used.

For a simple two-class problem with a positive class and a negative class using a linear kernel, a pattern x in the positive class gives $w^T x + b > 0$ and $w^T x + b < 0$ for x in the negative class. The positive and negative classes are characterized by two hyperplanes that are parallel such that the training patterns are near the respective hyperplane. These planes are called the support planes. The decision boundary or the separating plane is characterized by $w^T x + b = 0$ [15].

SVM models were trained. Four basic shapes were used for training a simpler model, named Model 1. Model 2 made use of data from seven shapes with multiple grasp variations. To evaluate these, confusion matrices were used so that a breakdown of the classification accuracy can be shown for each shape category.

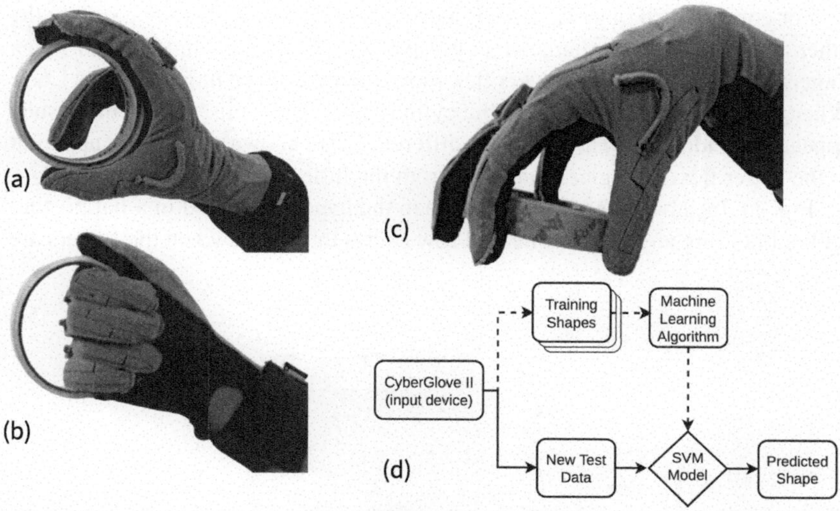

FIGURE 11.4 Sensing of global shape primitives through sensing of joint angles. (a-c) Different possible grasps for a ring (outside diameter = 90 mm, inside diameter = 75 mm) with a sensorized glove. Adapted from Anand et al. [13], with permission. ©2013 IEEE. (d) System architecture for the glove-based discrimination of shape primitives with the training and classification pathways.

11.3.2 Results

11.3.2.1 Confusion matrices

For Model 1, which consisted of a box, a cylinder, a flat surface, and a general grasp, the initial model was able to achieve a classification accuracy of 100% (Fig. 11.5a). The high score can be attributed to the simple shapes and a lack of grasp variation. The grasp required to perform the task was straightforward. To confirm the robustness of the results, an increased intra-class variability would be necessary.

For Model 2, there were seven shape categories that were used: a ball, box, cylinder, cone, flat surface, a ring, and a general grasp. The classification accuracy obtained was 71% (Fig. 11.5b). Misclassifications were apparent in the confusion matrix. The errors can be attributed to the challenges in distinguishing objects when multiple grasps are involved.

11.3.2.2 Shapes comparisons

Fig. 11.6a shows the results of the joint angle variations for the cylindrical materials. For the four sample objects (i.e., cola bottle, PET bottle, water bottle, and a small medicine bottle), the joint angles show similar patterns. The results from the thumb, middle, and ring fingers are significant in predicting a cylindrical shape. The results also indicate that the prediction of the shape does not depend on the size but on the shape of the object (i.e., cola bottle vs small medicine bottle).

For cone-shaped objects, there is a decreasing diameter from one side to the other. When the larger diameter is grasped by the thumb, middle, and index fingers, the ring and little fingers flex more than the other fingers (Fig. 11.6b). When compared to the results of the cylindrical shapes (Fig. 11.6a), the results appear to be identical. However, the differences are shown to be more apparent in the larger flexure angles than those from the little and ring fingers.

Fig. 11.7a shows the joint angles when the hand is oriented in a flat posture without holding any object. The first row shows the results when the fingers are

(a)

OUTPUT \ TARGET	Box	Cylinder	Flat	Grasp	SUM
Box	2 11.11%	0 0.00%	0 0.00%	0 0.00%	2 100.00% 0.00%
Cylinder	0 0.00%	11 61.11%	0 0.00%	0 0.00%	11 100.00% 0.00%
Flat	0 0.00%	0 0.00%	3 16.67%	0 0.00%	3 100.00% 0.00%
Grasp	0 0.00%	0 0.00%	0 0.00%	2 11.11%	2 100.00% 0.00%
SUM	2 100.00% 0.00%	11 100.00% 0.00%	3 100.00% 0.00%	2 100.00% 0.00%	18 / 18 100.00% 0.00%

(b)

OUTPUT \ TARGET	Sphere	Box	Cone	Cylinder	Flat	Grasp	Ring	SUM
Sphere	2 6.45%	0 0.00%	0 0.00%	1 3.23%	0 0.00%	0 0.00%	1 3.23%	4 50.00% 50.00%
Box	0 0.00%	2 6.45%	0 0.00%	0 0.00%	0 0.00%	0 0.00%	0 0.00%	2 100.00% 0.00%
Cone	0 0.00%	0 0.00%	2 6.45%	1 3.23%	0 0.00%	0 0.00%	0 0.00%	3 66.67% 33.33%
Cylinder	0 0.00%	0 0.00%	0 0.00%	5 16.13%	0 0.00%	0 0.00%	0 0.00%	5 100.00% 0.00%
Flat	0 0.00%	0 0.00%	0 0.00%	0 0.00%	3 9.68%	0 0.00%	0 0.00%	3 100.00% 0.00%
Grasp	0 0.00%	0 0.00%	0 0.00%	0 0.00%	0 0.00%	2 6.45%	0 0.00%	2 100.00% 0.00%
Ring	2 6.45%	0 0.00%	0 0.00%	4 12.90%	0 0.00%	0 0.00%	6 19.35%	12 50.00% 50.00%
SUM	4 50.00% 50.00%	2 100.00% 0.00%	2 100.00% 0.00%	11 45.45% 54.55%	3 100.00% 0.00%	2 100.00% 0.00%	7 85.71% 14.29%	22 / 31 70.97% 29.03%

FIGURE 11.5 Confusion matrices for comparing the models' predicted labels with the true labels for various grasp types. (a) Confusion matrix for Model 1. (b) Confusion matrix for Model 2. Reproduced from Anand et al. [13], with permission. ©2013 IEEE.

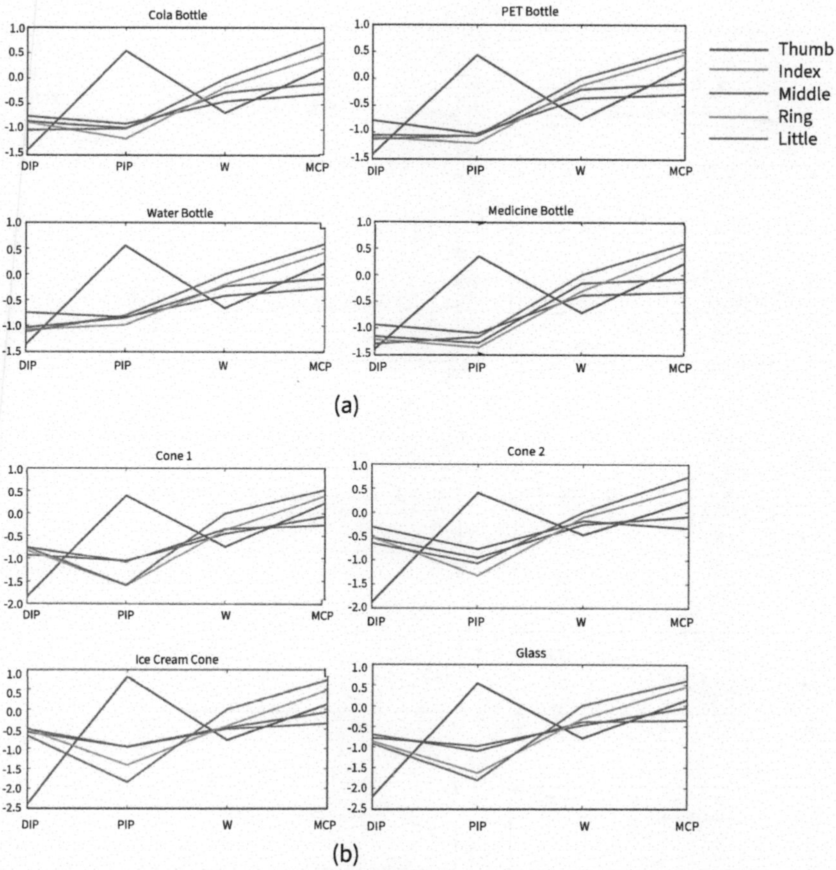

FIGURE 11.6 Results from the sensorized glove showing the relative angles taken from the different joints (i.e., distal interphalangeal (DIP), proximal interphalangeal (PIP), wrist (W), metacarpophalangeal (MCP) joints). Joint angles for (a) different cylinders and (b) for different cones. Reproduced from Anand et al. [13], with permission. ©2013 IEEE.

held closely together. The second row shows the joint angles when the fingers are held apart. The abduction/adduction posture of the hand resulted in the increase in joint angles primarily on the PIP joint of the thumb and the MCP joints of the other four fingers.

With reference to Fig. 11.4, the ring was held in three different types of grasps. Fig. 11.7b shows the joint angles for those grasps. When the ring was held in a bowling grasp (Fig. 11.4a), the thumb and the ring fingers are shown to be the distinguishing feature. When the ring was held similarly to Figs. 11.4b and 11.4c, the thumb, middle, and ring fingers are the more prominent features.

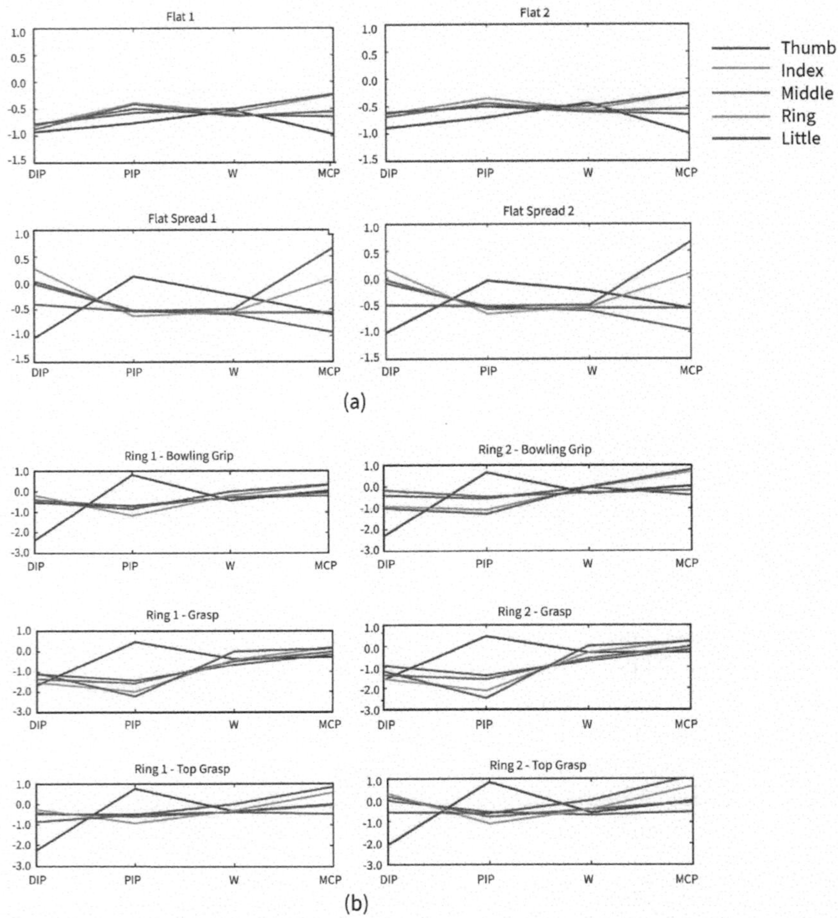

FIGURE 11.7 Results from the sensorized glove showing the relative angles taken from the different joints (i.e., distal interphalangeal (DIP), proximal interphalangeal (PIP), wrist (W), metacarpophalangeal (MCP) joints). Joint angles for (a) different flat finger postures and (b) different ring grasps. Reproduced from Anand et al. [13], with permission. ©2013 IEEE.

11.3.2.3 Fingers and joints comparisons

For both cone and cylinder grasps (Fig. 11.8a), the thumb exhibited similar joint angles. The index and middle fingers also follow similar patterns. The grasp on the cylinder, however, uses less joint flexion at the DIP and PIP joints. The ring and little fingers show significant differences. For grasping the cone, the fingers require larger flexion at the PIP joint. To differentiate a cone from the cylinder, large joint flexion will appear on the joints of the ring and little fingers for the cone when compared to the cylinder.

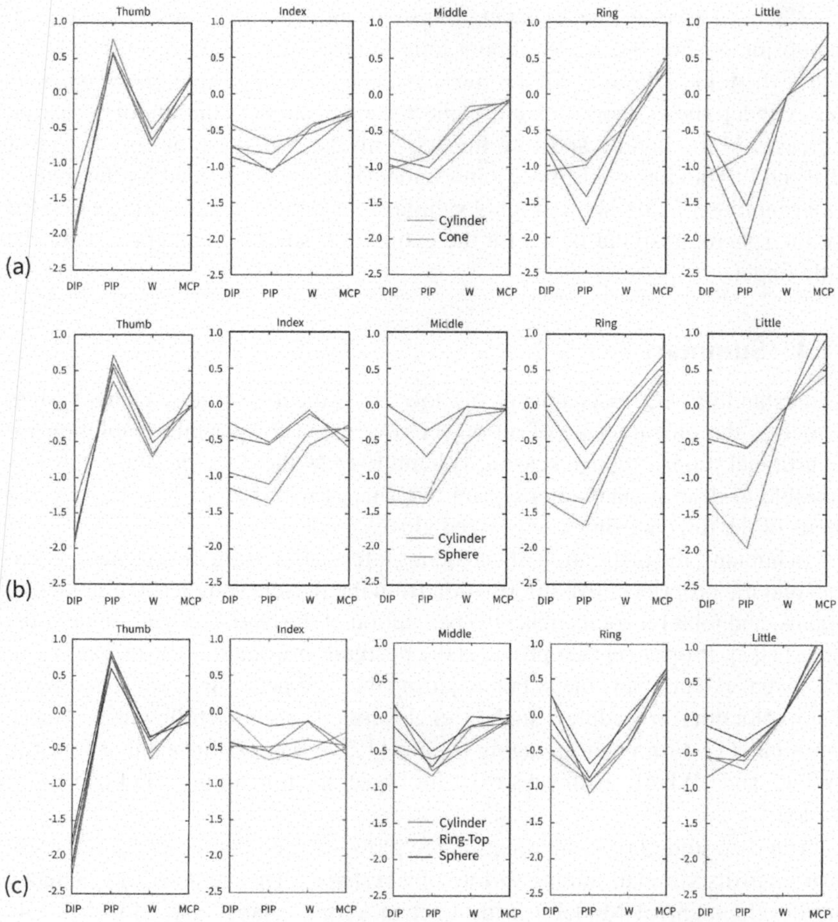

FIGURE 11.8 Results from the sensorized glove showing the relative angles taken from the different joints for each finger. Joint angles for (a) different cylinders and cones, (b) cylinders and spheres, and (c) sphere, cylinder, and top grasp of a ring object. Reproduced from Anand et al. [13], with permission. ©2013 IEEE.

When comparing cylinder and sphere grasps (Fig. 11.8b), the thumb shows only minor differences in the joint angles. The index finger exhibits greater flexion for the cylinder, particularly at the DIP and PIP joints, while MCP angles are similar. The little finger shows the largest differences in joint angles; specifically, grasping the cylinder requires increased DIP and PIP flexion compared to the sphere. The ring finger shows a similar pattern with the little finger. The index and middle fingers show highly significant differences at both DIP and PIP joints, where the cylinder requires much larger flexion when compared with the sphere.

When comparing the cylinder, ring-top, and sphere grasps (Fig. 11.8c), thumb joints show near-overlapping joint angles. The index finger shows the trend where the DIP and PIP are more extended for the sphere grasp, whereas the cylinder shows more flexion, while the ring-top is in the middle. Flexion increases in the middle and ring fingers, with the cylinder requiring the most DIP and PIP flexion, while these joints remain relatively extended for the sphere. The ring finger's DIP shows highly significant differences, remaining extended for the sphere grasp but flexed for the cylinder. A similar trend appears for the little finger.

11.4 Summary

Understanding the mechanisms of how the human tactile sensing system perceives the global shape of an object gave us an insight into a solution for an artificial proprioceptive sensing system. Even without tactile sensors, it was possible to discriminate among primitive shapes with the angle data from the joints of the fingers using a sensorized glove.

In humans, the integration of cutaneous information from the mechanoreceptors and the proprioceptive information from the muscle spindles, Golgi tendon organs, and joint receptors make the perception of the overall object shapes possible [1][2]. Proprioception provides the postural information corresponding to the spatial extent of an object. For acquiring the volumetric data, we also learned that exploratory procedures, such as enclosure and contour following, are also essential [11]. From neuroimaging studies [12], it was confirmed that separate "What" and "Where" pathways exist for object identification and object localization.

The case study involving a sensorized glove showed that proprioception can be used to discriminate among various object shapes. The methodology included the development of Model 1 with a set of simple shapes and Model 2 with another set of complex shapes and different grasp variations. Machine learning techniques were used to classify the finger joint angles that were obtained from the glove. Support Vector Machine with a radial basis function kernel [14] was found to be effective in classifying the shapes. The system achieved a 100% classification accuracy for the simple shapes and an accuracy of 71% for the complex shapes [13]. Results also showed that the ring and little fingers provide meaningful cues for differentiating between cylinders and cones.

In comparing a box vs. a cylinder, the distinguishing local surface features are the edges and vertices on a box, while a cylinder does not have those features. For a cone vs. a sphere, a cone is characterized by its changing curvature and tapered shape, while a sphere has constant curvature. For all those shapes, there are distinct variations in the joint angles of the fingers for a box vs. a cylinder, for a cone vs. a sphere, and for other pairwise comparison of primitive shapes.

Knowing the postural information from joint angles gave us an insight into the possibility of a proprioception-alone solution for artificial tactile sensing.

When complemented with tactile sensors on the fingertip, an artificial hand's sensing and perception of shapes can be improved. This has a potential to further improve the algorithms for reaching, preshaping, grasp planning, and overall control of an object.

References

[1] U. Proske, S.C. Gandevia, The proprioceptive senses: their roles in signaling body shape, body position and movement, and muscle force, Physiological Reviews 92 (2012) 1651–1697, https://doi.org/10.1152/physrev.00048.2011.

[2] S. Hsiao, Central mechanisms of tactile shape perception, Current Opinion in Neurobiology 18 (2008) 418–424, https://doi.org/10.1016/j.conb.2008.09.001.

[3] F. Solina, R. Bajcsy, Recovery of parametric models from range images: the case for superquadrics with global deformations, IEEE Transactions on Pattern Analysis and Machine Intelligence 12 (1990) 131–147, https://doi.org/10.1109/34.44401.

[4] J. Voisin, Y. Lamarre, C.E. Chapman, Haptic discrimination of object shape in humans: contribution of cutaneous and proprioceptive inputs, Experimental Brain Research 145 (2002) 251–260, https://doi.org/10.1007/s00221-002-1118-5.

[5] A.W. Goodwin, A.S. Browning, H.E. Wheat, Representation of curved surfaces in responses of mechanoreceptive afferent fibers innervating the monkey's fingerpad, Journal of Neuroscience 15 (1995) 798–810, https://doi.org/10.1523/JNEUROSCI.15-01-00798.1995.

[6] R.S. Johansson, I. Birznieks, First spikes in ensembles of human tactile afferents code complex spatial fingertip events, Nature Neuroscience 7 (2004 Jan) 170–177, https://doi.org/10.1038/nn1177.

[7] P. Jenmalm, I. Birznieks, A.W. Goodwin, R.S. Johansson, Influence of object shape on responses of human tactile afferents under conditions characteristic of manipulation, European Journal of Neuroscience 18 (2003) 164–176, https://doi.org/10.1046/j.1460-9568.2003.02721.x.

[8] J.C. Tuthill, E. Azim, Proprioception, Current Biology 28 (2018) R194–R203, https://doi.org/10.1016/j.cub.2018.01.064.

[9] J.J. Gibson, The Senses Considered as Perceptual Systems, Praeger, Westport, CT, 1983.

[10] G. Burton, M.T. Turvey, H.Y. Solomon, Can shape be perceived by dynamic touch?, Perception & Psychophysics 48 (1990) 477–487, https://doi.org/10.3758/bf03211592.

[11] S.J. Lederman, R.L. Klatzky, Hand movements: a window into haptic object recognition, Cognitive Psychology 19 (1987) 342–368, https://doi.org/10.1016/0010-0285(87)90008-9.

[12] C.L. Reed, R.L. Klatzky, E. Halgren, What vs. where in touch: an fMRI study, NeuroImage 25 (2005) 718–726, https://doi.org/10.1016/j.neuroimage.2004.11.044.

[13] A. Anand, J. Mathew, S. Pramod, S. Paul, R. Bharath, C. Xiang, J.J. Cabibihan, Object shape discrimination using sensorized glove, in: 2013 10th IEEE International Conference on Control and Automation (ICCA), 2013, pp. 1514–1519, https://doi.org/10.1109/icca.2013.6565154.

[14] C. Cortes, V. Vapnik, Support-vector networks, vol. 20, Machine Learning, 1995, pp. 273–297.

[15] B.E. Boser, I.M. Guyon, V.N. Vapnik, A training algorithm for optimal margin classifiers, in: Proceedings of the Fifth Annual Workshop on Computational Learning Theory (COLT '92), ACM, 1992, pp. 144–152, https://doi.org/10.1145/130385.130401.

Part IV

Structures

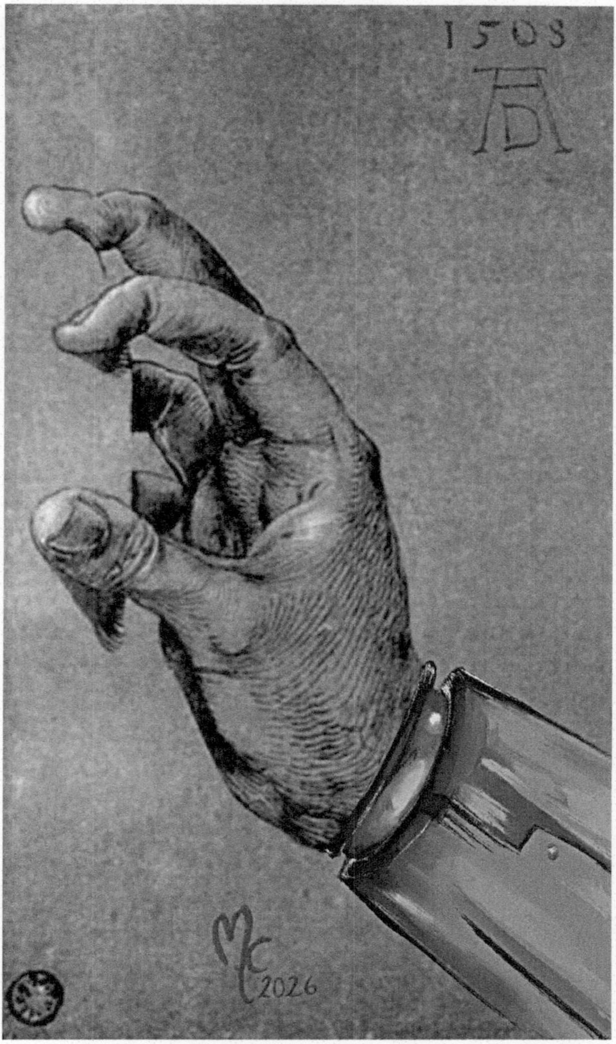

Chapter 12

Human and artificial hand structures

12.1 Introduction

The human hand affords us multiple actions that enable us to interact with and survive in the world. With our hands, we can gather and prepare food for ourselves, to feed ourselves and others. With our hands, we can bathe and clothe ourselves without the assistance of others. With our hands, we can go out, meet, and exchange social greetings with other people.

The human hand's dexterity is derived from the sequence of coordinated actions from the musculoskeletal system. This system consists of the bones, ligaments, tendons, and muscles (Fig. 12.1). In synchrony, the human hand can perform over 30 grasp poses, documented in the so-called grasp taxonomies [1][2][3]. The classification of the grasp postures differentiates between power and precision grasps, and further elaborates the nuances of each grasp according to an object's shape and size [4][5].

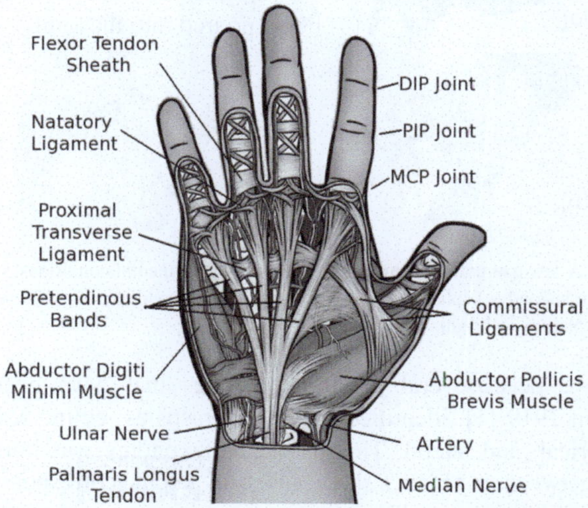

FIGURE 12.1 Depiction of the major tendons, ligaments, muscles, artery, and nerves at the volar side of the right hand. Reproduced from Cabibihan et al. [6]. ©2015 Cabibihan et al. This is an open access article distributed under the terms of the Creative Commons Attribution License (CC BY 4.0).

Humanizing Artificial Touch. https://doi.org/10.1016/B978-0-32-395519-5.00024-6

Inspired by the human hand, this chapter then describes some of the mechanical correlates that can help us design artificial hands. These include the structural, transmission, and actuation systems that have some parallelisms with the musculoskeletal system of the human hand. Structural systems are the parts and mechanisms constructed with stiff and soft structural materials that replicate the human hand's bones, ligaments, and cartilage [7][8]; transmission systems using cables and pulleys replicate the tendons [9][10]; and actuation systems with motors or pneumatic cylinders replicate the extrinsic or intrinsic muscles of the human hands [11][12].

This chapter has been subdivided into two parts: grasping with human hands (Sec. 12.2) and grasping with artificial hands (Sec. 12.4). Each subsection describes some of the important works on each topic. Within the subsection, the human hand's musculoskeletal system and the design elements of the artificial hands are described. Sec. 12.6 provides a summary of the chapter.

12.2 Grasping with human hands

To perform grasping or manipulation, the human hand relies on the synchronization of its anatomical parts to accomplish the desired task. The human grasps have been classified according to power grasps, precision grasps, and intermediate grasps (Fig. 12.2).

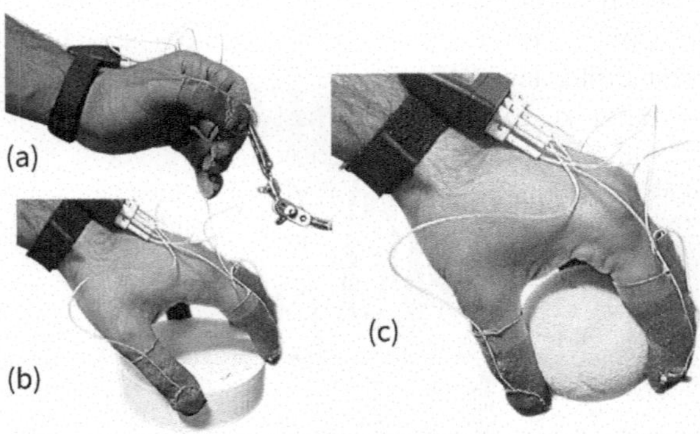

FIGURE 12.2 A few examples of human grasps under experimental conditions: (a) lateral pinch grasp, (b) disk grasp, and (c) spherical grasp. Adapted from Alhaddad et al. [13], with permission.

There have been several grasp taxonomies that have been proposed. Schlesinger in 1919 [1], identified the basic grasps as: cylindrical, tip, hook, palmar, spherical, and lateral. This particular taxonomy was born out of the relationship between an object's shape and size, and the corresponding type.

Napier's 1956 taxonomy [2] defined power grip as a grasp pose where an object is held in a clamp formed by the partly flexed fingers and the palm. He defined precision grip as a grasp where an object may be pinched between the fingers and the opposing thumb. In the 1960s, Landsmeer [14] gave further distinction between power and precision grasps. Power grasps, according to him, were more for stabilizing objects while precision grasps were meant for manipulating objects.

In addition to power and precision grasps, Kamakura et al. in the 1980s [3] defined intermediate grasps. These grasps are those grasps that have characteristics of both the power and precision grasps. In Kapandji's modes of prehension [15], he identified additional grasp types: cylindrical, digito-palmar, pollici-digital, lateral, and pentadigital grasps. Most of the taxonomies before him had a function-orientation in the grasp classifications. Kapandji showed the human grasp postures for affective and social postures. Those postures involve caresses and gestures.

Robotics researchers have built upon these earlier taxonomies and proposed some of the most comprehensive taxonomies that are currently being referenced in the design of robotic or prosthetic hands. In Cutkosky's taxonomy [4], which was partly developed by observing the grasps used by machinists as they work, he categorized the types of grasp according to the detailed task and geometry and contrasted those with gross task geometry. The taxonomy was further categorized according to a grasp's emphasis on stability in contrast to the emphasis on dexterity and sensibility.

More recently, the GRASP Taxonomy (Fig. 12.3; [5]) was able to synthesize all the previous grasp taxonomies. The grasps were categorized according to opposition type, their virtual finger assignments, power-intermediate-precision grasp type, and thumb position. All these result in a total of 33 unique prehensile grasp types that a human hand can do.

12.3 The human hand's musculoskeletal system

12.3.1 Bones

To reach and grasp objects with both arms, the upper extremities make use of 60 bones [16]. This accounts for 29% of the 206 total bones in the body. The bones at the upper extremities consist of the long bones of the humerus, ulna, and radius bones, and the bones at the hands (metacarpals and phalanges). The small bones include the bones at the wrist (carpals). To manipulate objects, the bones at both wrists (16 bones of the carpals) and hands (10 bones of the metacarpals and 28 bones of the phalanges) comprise 26% of the total bones in the body [15]. In other words, around a quarter of the bones in the body are needed to perform a bimanual manipulation task such as eating using cutlery or playing with a Rubik's cube.

Structurally, bones are considered as the rigid, non-deforming parts of the human body (Fig. 12.4). Bones are mostly composed of bone tissue, also known as osseous tissue [18]. This tissue is a type of connective tissue characterized by a collagen structure that is reinforced by calcium and phosphorus crystals. The bone tissue's capacity to preserve hardness without becoming brittle is due to the

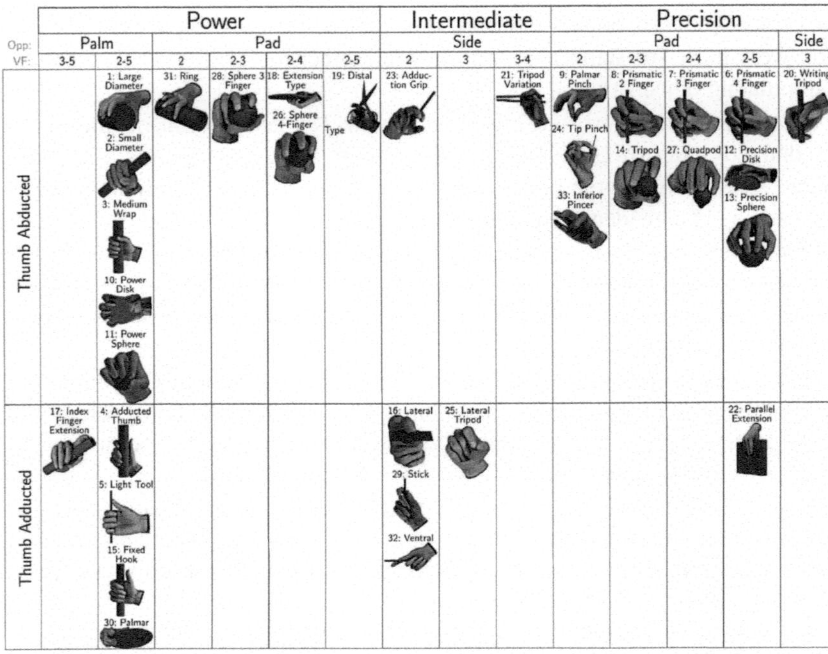

FIGURE 12.3 Taxonomy that includes previous grasp classifications from the literature. The grasps were categorized in the columns based on their classification into power, intermediate, and precision grasps, the type of opposition, and the virtual finger (VF) assignment. The rows were arranged according to the abducted or adducted postures of the thumb. Reproduced from Feix et al. [5], with permission. ©2016 IEEE.

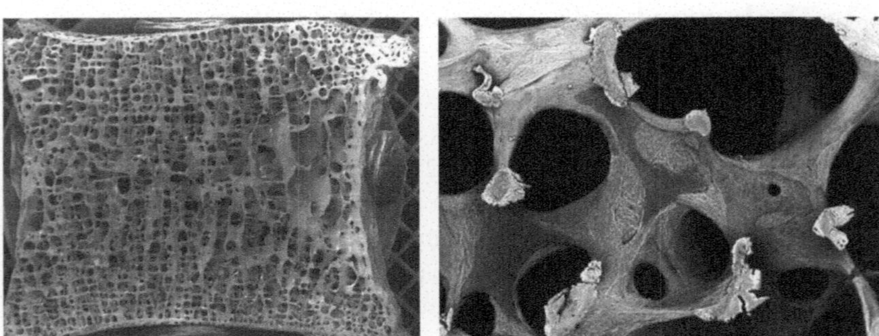

(a) (b)

FIGURE 12.4 Scanning Electron Microscopy (SEM) images of bone tissue. (a) Low magnification, wide field image of a 4 mm slice of L4 vertebra of a 30-year old female, reference scale: 2.7 mm grid shown in the background. (b) High magnification image of a sample L4 vertebra of a 61-year old male, scale: 500 μm. Reproduced from Boyde [17], with permission.

combination of a flexible collagen matrix that is mineralized with calcium and phosphorus crystals. The bones in the hand can withstand compressive stresses of around 170-190 MPa while their tensile strength is 50-150 MPa, which is lower than their compressive strength [19]. Bones are anisotropic, that is, the mechanical properties vary with the direction of the applied force; bones are stronger and stiffer along their longitudinal axis than their transverse axis [20].

Compact bone and spongy bone tissues are the two primary forms of osseous tissues [21]. The compact bone tissue makes up the external layer of the bone. This is also referred to as the hard or cortical bone. Its dense structure provides structural support and serves to protect the bone. The spongy bone tissue comprises the bone's internal structure. This is also referred to as the cancellous or trabecular bone. Its porous and honeycomb-like structure reduces the weight of the bone and provides load absorption capabilities.

12.3.2 Cartilage

The finger joints such as the metacarpophalangeal (MCP), proximal interphalangeal (PIP), and distal interphalangeal (DIP) joints are covered by articular cartilage. The articular cartilage in the fingers is located in between the phalanges of the fingers. It is a layer of low-friction, load bearing soft tissues that help achieve articulation, reduce friction, cushion impacts, and reduce stress concentrations during repetitive activities [22]. In their imaging studies, Saltzherr et al. [23] measured the thickness of the cartilage at the MCP joints to be around 0.3 and 0.9 mm and at the PIP joints to be around 0.2 and 0.7 mm. For those with osteoarthritis, a joint disease where the cartilage wears down causing bone-on-bone friction, the cartilage thickness was measured to be near 0.0 mm. Fig. 12.5 shows Magnetic Resonance Imaging (MRI) scans at the DIP of healthy and damaged cartilage at the DIP joint.

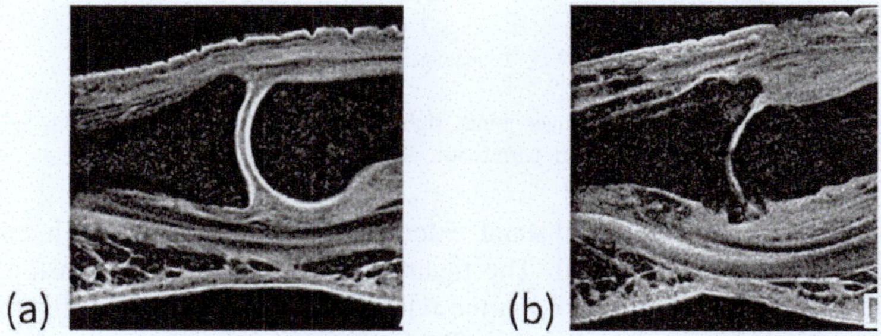

FIGURE 12.5 Magnetic Resonance Imaging (MRI) scan results of a DIP joint in different patients. (a) Healthy, no cartilage damage. (b) Severe cartilage damage with some bone-bone contact. Reproduced from Saltzherr [23], with permission.

In the study of Lu and Mow [22], they found that the cartilage exhibits elastic and viscoelastic behaviors. Its structure contributes to its tensile strength and compressive stiffness. The flow of interstitial fluid plays a crucial role in absorbing shocks and facilitating load transfer. The osmotic pressures generated by negatively charged glycosaminoglycans help maintain tissue hydration and resist compression. When cartilage is compressed, a volumetric change or a pressure gradient will be created within the tissue. That compression results in the interstitial fluid flowing within the tissue or extruding from the tissue. The fluid then passes through the porous solid matrix to generate high frictional resistance. This gives rise to the frictional dissipation resulting in viscoelastic behavior.

12.3.3 Ligaments

Ligaments serve the following functional roles: to provide joint stability, guide movements, and prevent excessive motions [24]. To provide joint stability, collateral ligaments can be found on both sides of the proximal interphalangeal (PIP) and distal interphalangeal (DIP) joints to prevent dislocation and to provide stability (Fig. 12.6). On the palm side of the PIP and DIP joints are the volar plates. These prevent the hyperextension of the joints. To guide finger movements, the annular and cruciate pulley ligaments provide a fiber and bone-type of structure as a tunnel for the flexor tendons. This functions to keep the flexor tendons close to the bone during flexion and extension. To prevent excessive motions, the interosseous ligaments between the metacarpal bones prevent the excessive spread or collapse of the hand.

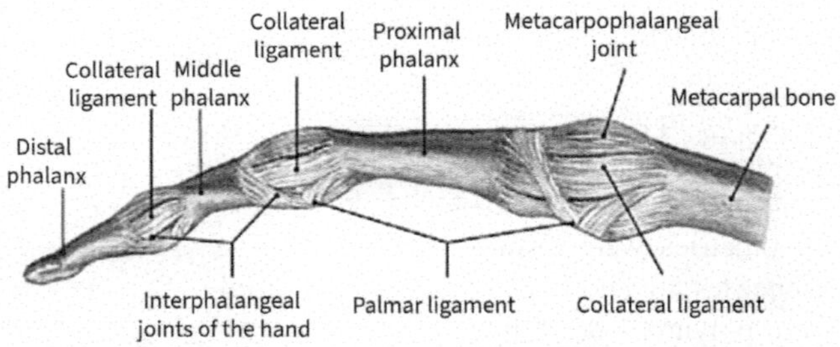

FIGURE 12.6 Ligaments of the finger joints, right side; lateral view. Adapted from Sobotta-Archive of Paulsen, Waschke [25], with permission. ©2022 Elsevier GmbH.

The human hand can withstand external forces due to the high tensile strength of the ligaments [26]. The ligaments can stretch under tension and return to their original length upon the release of forces. Moreover, ligaments absorb impact and prevent injuries. The stiffness of the ligaments depends on their functional role; collateral ligaments are stiffer because they need to stabilize the joints while the other ligaments are more compliant because they are needed for allowing a larger range of motion. Ligaments exhibit viscoelastic behavior, which can help adapt to various loads and reduce injuries.

Nordin and Frankel [24] describe ligaments to be composed of several components. The primary structural component of a ligament is the type

1 collagen. This component provides high tensile strength and resistance to overstretching. There is also a small amount of type 3 collagen, which contributes to the ligament's structural integrity and self-healing properties. Interspersed among the collagen fibers are the elastin fibers. These provide elasticity and the ability to make the ligament return to its original shape. Similar to the cartilage, ligaments also have proteoglycans. These help the extracellular matrix to collect and retain water, thus enabling the ligament to increase its viscoelastic properties and provide compressive strength. Ligaments have 60-70% water content by weight.

12.3.4 Tendons

Tendons in the human hand act as cables, connecting muscles and the bones, to transmit forces. For the movement of the hand and fingers to occur, tendons transmit the forces generated by the muscles to the bones, which allow grasping tasks of daily living. The fingers' flexion and extension movements are due to specific tendons (Fig. 12.7). The flexor tendons (Flexor Digitorum Superficialis and Flexor Digitorum Profundus) make fingers bend or curl while the extensor tendons (Extensor Digitorum) are responsible for making the fingers straight [27]. For fine manipulation and control, such as playing the piano or guitar, the tendons allow such precise movements to be possible [28].

Some of the important mechanical properties of tendons include: high tensile strength, elasticity, and viscoelasticity. The tendon's high tensile strength allows it to withstand excessive pulling forces especially when the muscles are in contraction [29]. Tendons have shape-recovery properties due to their elasticity. Tendons can stretch yet return to their original state after being stretched [30]. During dynamic activities, the tendon's viscoelastic properties allow it to absorb and dissipate energy to provide the needed protection for the muscles and bones during dynamic and repetitive grasping tasks.

12.3.5 Muscles

The muscles in the human hand are primarily responsible for the power and precision grasps used in daily life. The muscles that make the movements of the hand possible are categorized as extrinsic and intrinsic muscles. Extrinsic muscles of the hand originate in the forearm and extend into the hand via tendons, which attach to the bones of the wrist, hand, and fingers [31]. The extrinsic muscles are responsible for the gross movements of the wrist and fingers. The flexor muscles allow the middle phalanges of the fingers to flex (Flexor Digitorum Superficialis; [32]). The Flexor Digitorum Profundus allows the distal phalanges to flex [33]. For the thumb, the muscle responsible to flex it is the Flexor Pollicis Longus [34].

Intrinsic muscles are within the hand (Fig. 12.8). These muscles are important for the precise movements of the fingers. The thenar muscles,

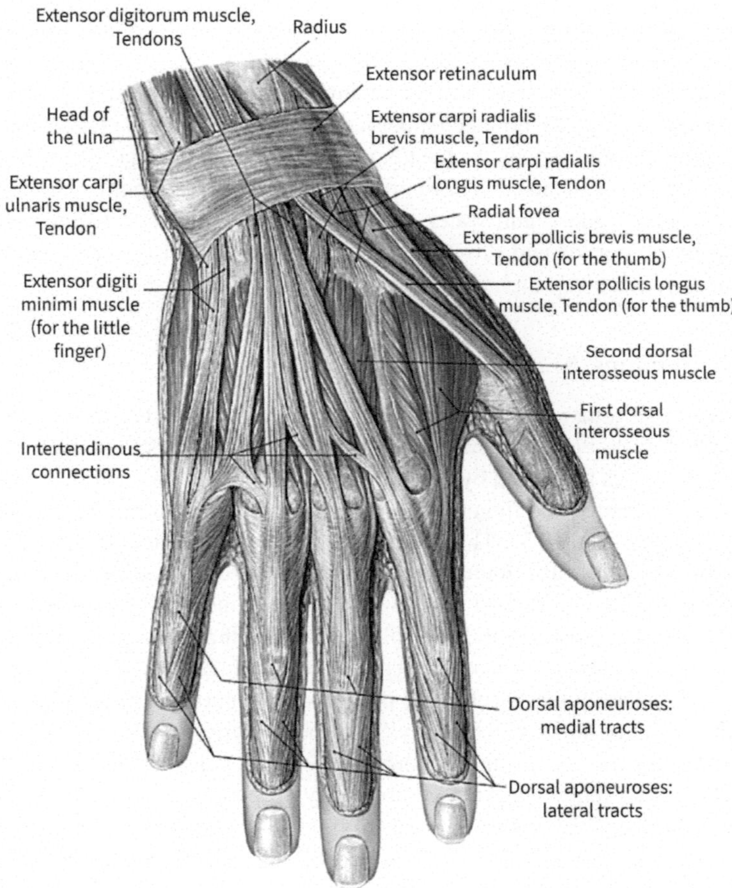

FIGURE 12.7 Tendons of the back of the hand, right side; dorsal view. Adapted from Sobotta-Archive of Paulsen, Waschke [25], with permission. ©2022 Elsevier GmbH.

classified as intrinsic muscles, perform three main movements [35]: abduction of the thumb (Abductor Pollicis Brevis), flexion of the thumb (Flexor Pollicis Brevis), and opposition of the thumb to the other fingers (Opponens Pollicis).

The lumbrical muscles have the function of flexing the metacarpophalangeal (MCP) joints and lengthening the interphalangeal joints [36]. The dorsal interosseous muscles contribute to finger abduction, while the palmar interosseous muscles contribute to finger adduction [37].

The mechanical properties of the muscles in the human hand enable it to perform the various power and precision grasps. Some of the key properties are their force generation capability, stiffness, and elasticity. In the studies of Chen et al. [38] on the strength of the hands of adolescents and young adults, they showed that intrinsic muscles can reach a maximum force of up to 149 N for male participants and up to 97 N for female participants.

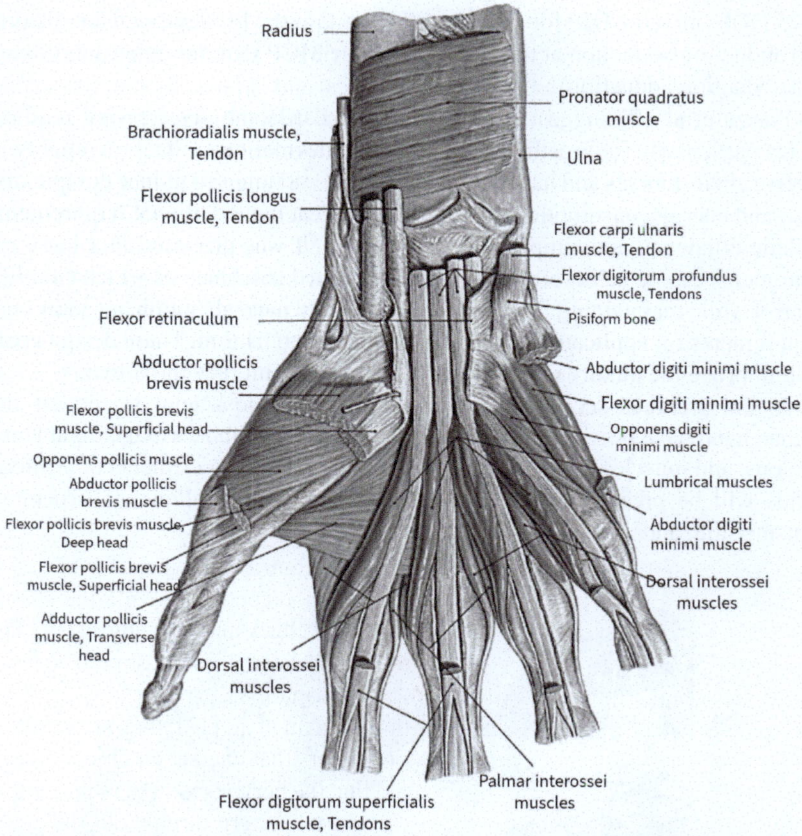

FIGURE 12.8 Muscles in the palm of the hand, right side; palmar view; after the removal of the palmar aponeurosis and the superficial muscles. Adapted from Sobotta-Archive of Paulsen, Waschke [25], with permission. ©2022 Elsevier GmbH.

12.4 Grasping with artificial hands

What design requirements must artificial hands meet to successfully perform grasping tasks? Melchiorri and Kaneko [11] suggest that both anthropomorphism and dexterity are crucial requirements that need to be addressed by designers of artificial hands. Anthropomorphism implies that both appearance and function of the human hands need to be replicated [39]. Anthropomorphic features include multiple fingers and an opposable thumb.

Dexterity then lets the hand execute a diverse set of tasks. This includes tasks involving delicate precision grips to powerful ones. The human hand's dexterity can be attributed to its intricate structure that includes a combination of bones, muscles, and tendons that work to enable fine motor control and strength. The human hand has 21 degrees of freedom (DoF). There are 4 DOFs in each of the 4 fingers (3 for extension/flexion and 1 for abduction/adduction). The thumb has 5 DOFs (1 at the DIP for extension/flexion, 2 at the PIP for extension/flexion and abduction/adduction at the PIP, and 2 at the MCP joint for extension/flexion and abduction/adduction).

Piazza et al. [7] suggest that robustness and adaptability are vital requirements particularly for prosthetic devices. Such devices must function reliably in diverse environments and handle unexpected interactions. To reduce complexity, cost, and energy consumption, they further suggest that additional requirements include efficiency and simplicity in the design. It was proposed that these requirements can be achieved through underactuated mechanisms which simplify control while maintaining functionality. Safety and natural motion are additional requirements for applications that involve human interaction. These design goals help improve the quality of life of the users and the interaction partner.

In the previous section, elements of the musculoskeletal system of the human hand have been presented. This included the bones, cartilage, ligaments, tendons, and muscles. In the following sections, the design elements of artificial hands will be presented (Fig. 12.9). This includes the following: structural, transmission, and actuation systems.

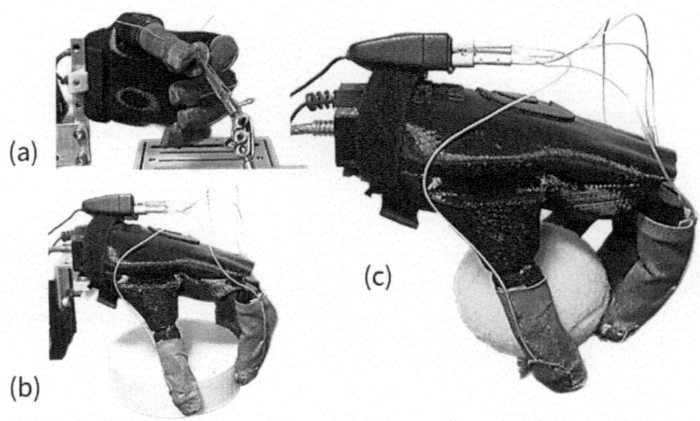

FIGURE 12.9 A few examples of grasps with a 3D printed artificial hand under experimental conditions: (a) lateral pinch grasp, (b) disk grasp, and (c) spherical grasp. Adapted from Alhaddad et al. [13], with permission. ©2017 Springer Nature. All rights reserved.

12.5 Design elements of artificial hands

12.5.1 Structural system

In the field of buildings and construction, the term structural system refers to the load-resisting elements, which include tensile structures, compressive structures, trusses, frames, buttresses, among others. In the musculoskeletal system, these elements are analogous to the bones, cartilage, and ligaments. In the design of artificial hands, these structural elements known in human hands have been lumped into topics such as joints, linkages, and kinematics. Likewise, for this chapter, the bones, cartilage, and ligaments will be lumped as the structural system of the artificial hand.

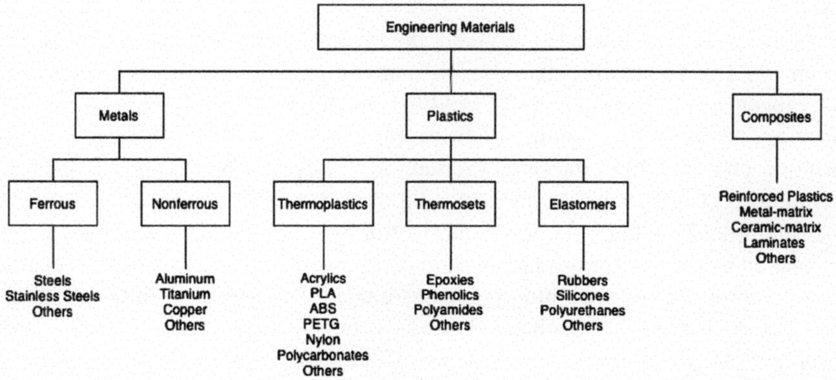

FIGURE 12.10 Taxonomic classification of engineering materials with particular usage in the construction of the structures of artificial hands. The typical materials used are metals, plastics, and composites.

In Piazza et al. [7], they evaluated the majority of the scientific articles on robotic hands from 1912-2018. They summarized the robot hand structures according to their joints as being rigid, flexible, dislocatable, and soft continuous. Finger designs with rigid joints have links that are connected using fixed mechanical elements, such as pins. This was the typical design used in many early stages of robot hand designs. Fingers with flexible joints are finger structures where the links are connected using flexible elements, such as springs. This permits more compliance and adaptability in the hand's movements. Fingers with dislocatable joints are finger structures where the links are connected using elastic and flexible elements that can withstand severe disarticulations. This was said to have even better flexibility and adaptability as compared to rigid or flexible joints. More recently, designs with fingers that have soft continuous joints are being designed using continuously flexible materials, rather than discrete links and joints. This allowed the soft fingers to conform to objects in a very compliant and adaptive manner.

Various engineering materials have been used to construct the structural systems of artificial hands. In Fig. 12.10, both ferrous and non-ferrous metals were commonly used in the past. In addition, plastics and composites have been used due to their advantageous mechanical properties. Metals are used for their strength and high stiffness. Steel or aluminum is suitable as the mechanical structure and components of artificial hands. The DLR Hand-II's structure made use of aluminum to achieve high dexterity and accuracy in the manipulation of objects [40]. For the finger joints and outer casings, plastics were used due to their lightweight and moldable characteristics [41].

Composite materials have also been used because they possess the combined benefits of strength and weight. The Cyberhand, for example, had a composite material as the structural component of the robot hand's palm [42]. Composites enhance the mechanical properties of robotic hands while reducing overall mass.

Soft materials are needed to create the skins of the robot's fingers to adapt to an object's shape. To create flexible and compliant components, elastomers have been used in the past [43]. Additive manufacturing, also known as 3D printing, has become a common fabrication method for robotic hands. The 3D printing process allows the rapid production of a part or for creating free-form geometries [44][8].

Traditional methods for metal fabrication (e.g., machining, forming) were employed especially for producing rigid components that require high precision and durability [44]. Machining processes enable the creation of intricate parts that can withstand the mechanical stresses. For gears, mechanism links, and structural supports used in robot hands, Computer Numerical Control (CNC) machining is a common process to manufacture metal parts.

For many centuries, parts were formed by casting metal to form an arbitrary shape. A similar technique is being used for casting soft, silicone materials. In this process, liquid material is poured into molds to create specific shapes. The material is allowed to cure for a few hours to solidify. In small batch quantities for prototyping, the molds are constructed with the 3D printing process [45].

12.5.2 Transmission system

The transmission system is responsible for the transfer of movement and power from the actuator to the fingers so that the grasping tasks can commence [11]. In the human hand, this role is accomplished by the tendon. The conventional transmission systems documented in the literature encompass tendon-driven mechanisms, gear-based systems, and hybrid designs [39].

Gear-driven transmission systems offer greater torque and exact control over finger movements. Functionally, they utilize a combination of gears and links to convert motor rotations into joint movements. The DLR/HIT Hand employs a gear train to provide intricate finger movements to allow dexterous manipulation [46]. The gears at the extremities of the finger structures are similar to the cartilage located at the ends of the finger phalange bones. Gear systems offer strong performance but they are generally larger and bulkier than tendon-driven versions. Consequently, this potential restricts their use in lightweight mechanical hands.

Cable-driven, or tendon-driven devices, are now more commonly used as the transmission system of artificial hand designs due to their lightweight and flexible characteristics. These emulate the biological tendons of the human hand to facilitate a more natural range of motion. To flex and extend the fingers, cables are routed through cavities in the artificial hand's structure to connect motors to the joints of the fingers [10][39].

Those cavities are analogous to the ligaments in the human hand. However, unlike biological ligaments, these artificial channels are susceptible to friction and wear, which may affect the precision of movements over time [47]. Some of the initial works that utilized tendon-driven mechanisms are the Stanford/JPL hand and the Utah/MIT dexterous hand [48].

12.5.3 Actuation system

To generate the movements and forces for artificial hands, an actuation system is necessary. Birglen et al. [50] credits the dexterity of an artificial hand to the actuators. Melchiorri and Kaneko [11] categorized actuation methods in artificial hands according to their positioning within the robotic hand: remote and in-situ actuation. In remote actuation, the actuators are positioned at a distance from the hand. Typically, this location is at the forearm where tendon cables transmit the movement to the fingers. In-situ actuation refers to the placement of actuators within the hand. This has an advantage of more compact designs and potentially more responsive control. The categories originated from the extrinsic and intrinsic muscles of the human arm and forearm [51].

In humans, extrinsic muscles, like the remote actuators, originate from the forearm by acting on the tendons that extend into the hand. Intrinsic muscles, like the in-situ actuators, are located within the hand itself. Comparisons on the various characteristics of the human muscles and other types of actuators are given in Figs. 12.11 and 12.12.

Melchiorri and Kaneko [11] state that if we consider $\boldsymbol{M}$ to be the number of actuators that directly or indirectly drive the joints in the artificial fingers and $\boldsymbol{N}$ to be the overall number of joints, we will obtain another categorization of actuation architecture. In the literature, the most common architectures include fully actuated, underactuated, and coupled systems [11]. For an $\boldsymbol{M} = \boldsymbol{N}$ architecture, the number of actuators $\boldsymbol{M}$ equals the number of joints $\boldsymbol{N}$ of artificial fingers. These are known as fully actuated artificial hand designs.

A high degree of dexterity can be achieved by having multiple joints that can be controlled by individual actuators. For $\boldsymbol{M} < \boldsymbol{N}$, the number of actuators is fewer than the number of joints. These are known as underactuated systems. These allow a simpler design while still achieving a range of grasping capabilities. This architecture makes use of passive elements, such as springs, to enable compliant movements and adapt to the shape of the objects being grasped. For $\boldsymbol{M} > \boldsymbol{N}$, the number of actuators exceeds the number of joints. This architecture involves having more than one actuator operating on a single joint. Such an approach can simplify the design but has the drawback of limited precision in the movements.

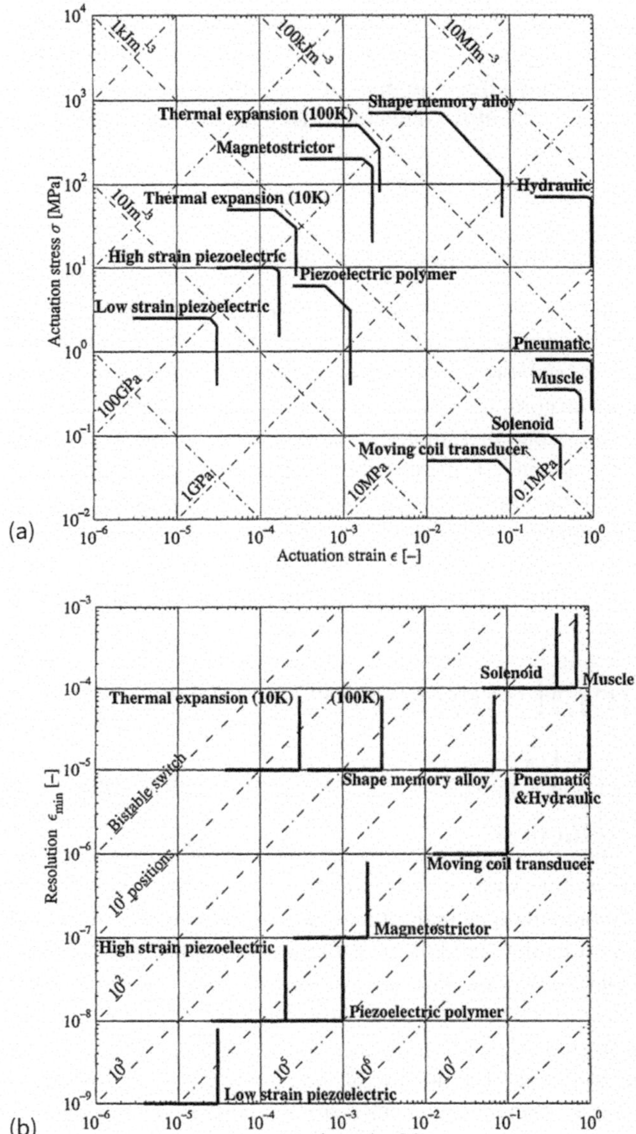

FIGURE 12.11 Characteristics of mechanical actuators. (a) Actuation stress, σ, versus actuation strain, ϵ. Bold lines denote the upper limits of actuation performance. (b) Strain resolution, ϵ_{min}, versus the actuation strain, ϵ, for various actuators. Bold lines denote the limits of actuation performance. Reproduced from Huber et al. [49], with permission. ©1997 The Royal Society (UK). All rights reserved.

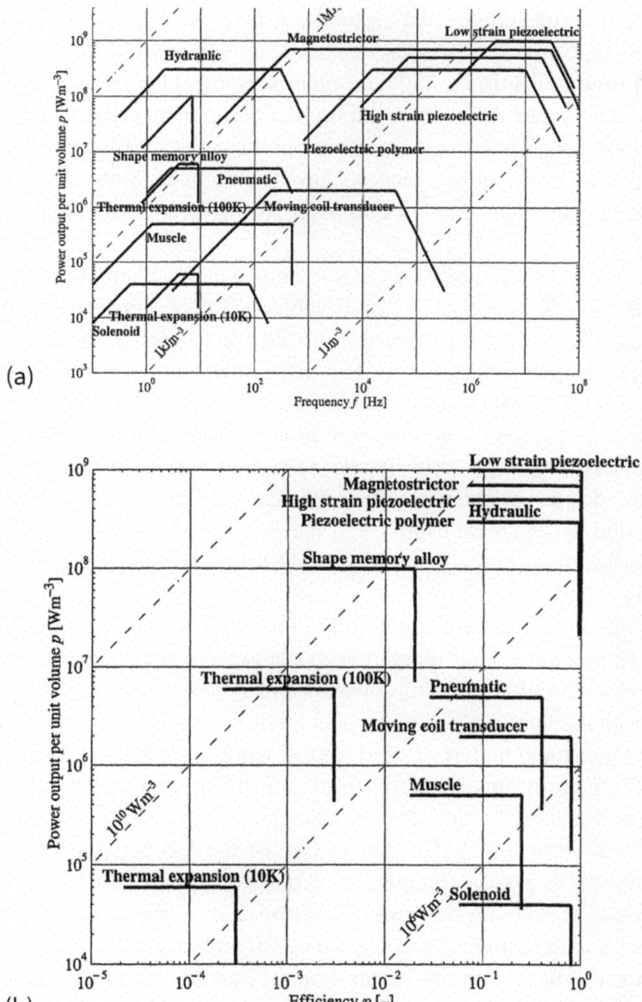

FIGURE 12.12 Characteristics of mechanical actuators. (a) Volumetric power, p, versus frequency, f, for various actuators. (b) Volumetric power, p, versus efficiency, ν, for various actuators. Reproduced from Huber et al. [49], with permission.

12.6 Summary

Understanding the human hand's musculoskeletal system and the intricate coordination of bones, cartilage, ligaments, tendons, and muscles will allow us to humanize the structure of the artificial hand of humanoid robots and prosthetics.

The upper extremities have a total of 60 bones and have a complex arrangement of connective tissues to afford us with a wide range of grasping tasks, from powerful grips to delicate manipulations [15]. From the studies of the mechanical properties of these biological components that were presented, we found that bones have anisotropic properties, the cartilage and ligaments have viscoelastic behavior, the tendons have high tensile strength and elasticity, and muscles are situated in two locations to allow us to perform gross and fine movements.

The taxonomy of human grasps was reviewed in this chapter where the early classifications by Schlesinger [1] and Napier [2] were presented and we saw how this taxonomy evolved from a more detailed classification as presented in Cutkosky [4] and in Feix et al. [5]. It is now common to see designs of artificial hands being benchmarked to these taxonomies. The GRASP taxonomy, for example, identifies 33 different grasp types, which can be used to assess how well a robotic hand can perform specific tasks that require dexterity.

The key design goals were then presented. For effective human-robot interaction and for practical usability, it was suggested that anthropomorphism, dexterity, robustness, and adaptability of an artificial hand's design should be considered [11][7].

Borrowing from the human hand's musculoskeletal system, analogous structural, transmission, and actuation systems were presented. To mimic the human hand's structural system, consisting of bones, cartilage, and ligaments, some of the characteristics of metals, plastics, and composites were described as these were the typical materials used in past works, which were then processed through 3D printing and CNC machining [8][44]. Inspired by the tendons, transmission systems in artificial hands employ gear-based, tendon-driven, or hybrid mechanisms [9][10]. Analogous to the human muscles, actuation systems employ DC motors, pneumatic actuators, and novel materials like shape memory alloys. These are designed to optimize dexterity and control through fully actuated, underactuated, or coupled configurations [50].

Knowing the basic aspects of the human musculoskeletal system that are related to the grasping and manipulation of objects serves as a useful guide in the design and fabrication of artificial hand structures that can replicate some of the functional and social aspects of the human hand.

References

[1] G. Schlesinger, Der mechanische Aufbau der künstlichen Glieder. Ersatzglieder und Arbeitshilfen, Springer, Berlin, Heidelberg, 1919, pp. 321–661, https://doi.org/10.1007/978-3-662-33009-8_13.

[2] J.R. Napier, The prehensile movements of the human hand, Journal of Bone and Joint Surgery. British Volume 38-B (1956) 902–913, https://doi.org/10.1302/0301-620x.38b4.902.

[3] N. Kamakura, M. Matsuo, H. Ishii, F. Mitsuboshi, Y. Miura, Patterns of static prehension in normal hands, The American Journal of Occupational Therapy 34 (1980 Jul) 437–445, https://doi.org/10.5014/ajot.34.7.437.

[4] M. Cutkosky, On grasp choice, grasp models, and the design of hands for manufacturing tasks, IEEE Transactions on Robotics and Automation 5 (1989 Jun) 269–279, https://doi.org/10.1109/70.34763.

[5] T. Feix, J. Romero, H. Schmiedmayer, A.M. Dollar, D. Kragic, The GRASP taxonomy of human grasp types, IEEE Transactions on Human-Machine Systems 46 (2016) 66–77, https://doi.org/10.1109/THMS.2015.2470657.

[6] J.J. Cabibihan, D. Joshi, Y.M. Srinivasa, M.A. Chan, A. Muruganantham, Illusory sense of human touch from a warm and soft artificial hand, IEEE Transactions on Neural Systems and Rehabilitation Engineering 23 (2015) 517–527, https://doi.org/10.1109/tnsre.2014.2360533.

[7] C. Piazza, G. Grioli, M. Catalano, A. Bicchi, A century of robotic hands, Annual Review of Control, Robotics, and Autonomous Systems 2 (2019 May) 1–32, https://doi.org/10.1146/annurev-control-060117-105003.

[8] R.N.M. Delda, R.B. Basuel, R.P. Hacla, D.W.C. Martinez, J.J. Cabibihan, J.R.C. Dizon, 3D printing polymeric materials for robots with embedded systems, Technologies 9 (2021 Nov) 82, https://doi.org/10.3390/technologies9040082.

[9] J.T. Belter, J.L. Segil, A.M. Dollar, R.F. Weir, Mechanical design and performance specifications of anthropomorphic prosthetic hands: a review, Journal of Rehabilitation Research and Development 50 (2013) 599, https://doi.org/10.1682/jrrd.2011.10.0188.

[10] C. Cipriani, M. Controzzi, M.C. Carrozza, The SmartHand transradial prosthesis, Journal of NeuroEngineering and Rehabilitation 8 (2011) 29, https://doi.org/10.1186/1743-0003-8-29.

[11] C. Melchiorri, M. Kaneko, Robot hands, in: Springer Handbook of Robotics, Springer, Berlin, Heidelberg, 2008, pp. 345–360.

[12] J. Szkopek, G. Redlarski, Artificial-hand technology—current state of knowledge in designing and forecasting changes, Applied Sciences 9 (2019 Sep) 4090, https://doi.org/10.3390/app9194090.

[13] A.Y. Alhaddad, S.E. AlKhatib, R.A. Khan, S.M. Ismail, A.S.S. Shehadeh, A.M. Sadeq, J.J. Cabibihan, Toward 3D printed prosthetic hands that can satisfy psychosocial needs: grasping force comparisons between a prosthetic hand and human hands, in: A. Kheddar, E. Yoshida, S.S. Ge, K. Suzuki, J.J. Cabibihan, F. Eyssel, H. He (Eds.), Social Robotics, Springer International Publishing, Cham, 2017, pp. 304–313, https://doi.org/10.1007/978-3-319-70022-9_30.

[14] J. Landsmeer, Power grip and precision handling, Annals of the Rheumatic Diseases 21 (1962 Jun) 164–170, https://doi.org/10.1136/ard.21.2.164.

[15] I.A. Kapandji, The Physiology of the Joints: Upper Limb, vol. 1, 5th edn, Churchill Livingstone, London, England, 1982.

[16] A. Tozeren, Human Body Dynamics, 2000th ed., Springer, New York, NY, 2000 Dec.

[17] A. Boyde, Scanning electron microscopy and bone, in: Bone Research Protocols, Springer, US, 2025, pp. 621–670, https://doi.org/10.1007/978-1-0716-4306-8_31.

[18] S. Weiner, H.D. Wagner, The material bone: structure-mechanical function relations, Annual Review of Materials Science 28 (1998 Aug) 271–298, https://doi.org/10.1146/annurev.matsci.28.1.271.

[19] J.Y. Rho, L. Kuhn-Spearing, P. Zioupos, Mechanical properties and the hierarchical structure of bone, Medical Engineering & Physics 20 (1998 Mar) 92–102, https://doi.org/10.1016/s1350-4533(98)00007-1.

[20] J.D. Currey, Bones: Structure and Mechanics, Princeton University Press, Princeton, NJ, 2002 Jun.

[21] J.A. Buckwalter, M.J. Glimcher, R.R. Cooper, R. Recker, Bone biology, The Journal of Bone and Joint Surgery 77 (1995 Aug) 1256–1275, https://doi.org/10.2106/00004623-199508000-00019.

[22] X.L. Lu, V.C. Mow, Biomechanics of articular cartilage and determination of material properties, Medicine & Science in Sports & Exercise 40 (2008) 193–199, https://doi.org/10.1249/mss.0b013e31815cb1fc.

[23] M. Saltzherr, G. Muradin, I. Haugen, R. Selles, J. van Neck, J. Coert, J. Hazes, J. Luime, Cartilage evaluation in finger joints in healthy controls and early hand osteoarthritis patients using high-resolution MRI, Osteoarthritis and Cartilage 27 (2019) 1148–1151, https://doi.org/10.1016/j.joca.2019.05.003.
[24] M. Nordin, V. Frankel, Basic Biomechanics of the Musculoskeletal System, Lippincott Williams and Wilkins, Philadelphia, PA, 2020.
[25] J. Waschke, F. Paulsen (Eds.), Sobotta Atlas der Anatomie, Elsevier, Munchen, Germany, 2022, 25th edition.
[26] S.L.Y. Woo, R.E. Debski, J. Zeminski, S.S. Chan Saw, J.A. Fenwick, Injury and repair of ligaments and tendons, Annual Review of Biomedical Engineering 2 (2000) 83–118, https://doi.org/10.1146/annurev.bioeng.2.1.83.
[27] H.J. Goodman, J. Choueka, Biomechanics of the flexor tendons, Hand Clinics 21 (2005) 129–149, https://doi.org/10.1016/j.hcl.2004.11.002.
[28] P.L. Kuo, D.L. Lee, D.L. Jindrich, J.T. Dennerlein, Finger joint coordination during tapping, Journal of Biomechanics 39 (2006) 2934–2942, https://doi.org/10.1016/j.jbiomech.2005.10.028.
[29] J.H.C. Wang, Mechanobiology of tendon, Journal of Biomechanics 39 (2006) 1563–1582, https://doi.org/10.1016/j.jbiomech.2005.05.011.
[30] P. Kannus, Structure of the tendon connective tissue, Scandinavian Journal of Medicine & Science in Sports 10 (2000) 312–320, https://doi.org/10.1034/j.1600-0838.2000.010006312.x.
[31] M. Mirakhorlo, J.M.A. Visser, B.A.A.X. Goislard de Monsabert, F.C.T. Helm, H. van der Maas, H.E.J. Veeger, Anatomical parameters for musculoskeletal modeling of the hand and wrist, International Biomechanics 3 (2016) 40–49, https://doi.org/10.1080/23335432.2016.1191373.
[32] P. Yoong, R. Goodwin, A. Chojnowski, Phalangeal fractures of the hand, Clinical Radiology 65 (2010) 773–780, https://doi.org/10.1016/j.crad.2010.04.008.
[33] K.J. Chepla, R.J. Goitz, J.R. Fowler, Anatomy of the flexor digitorum profundus insertion, The Journal of Hand Surgery 40 (2015) 240–244, https://doi.org/10.1016/j.jhsa.2014.11.004.
[34] L.L. Bruin, J. Lans, K.R. Eberlin, N.C. Chen, Zone II flexor pollicis longus repair: thumb flexion and complications, HAND 17 (2020) 285–292, https://doi.org/10.1177/1558944720930294.
[35] S. Gupta, H. Michelsen-Jost, Anatomy and function of the thenar muscles, Hand Clinics 28 (2012) 1–7, https://doi.org/10.1016/j.hcl.2011.09.006.
[36] R. Palti, M. Vigler, Anatomy and function of lumbrical muscles, Hand Clinics 28 (2012) 13–17, https://doi.org/10.1016/j.hcl.2011.09.002.
[37] Y.K. De Almeida, M. Krebs, M. Braun, F. Dap, G. Dautel, L. Athlani, Innervation and vascular supply of the first dorsal interosseous muscle and palmar interosseous muscle of the index: an anatomic descriptive study, Morphologie 105 (2021) 298–307, https://doi.org/10.1016/j.morpho.2020.12.008.
[38] C.Y. Chen, C.W. McGee, T.L. Rich, C.N. Prudente, B.T. Gillick, Reference values of intrinsic muscle strength of the hand of adolescents and young adults, Journal of Hand Therapy 31 (2018) 348–356, https://doi.org/10.1016/j.jht.2017.05.012.
[39] M. Controzzi, C. Cipriani, M.C. Carrozza, Design of Artificial Hands: A Review. The Human Hand as an Inspiration for Robot Hand Development, Springer International Publishing, 2014, pp. 219–246, https://doi.org/10.1007/978-3-319-03017-3_11.
[40] J. Butterfass, M. Grebenstein, H. Liu, G. Hirzinger, DLR-Hand II: Next Generation of a Dextrous Robot Hand, Proceedings 2001 ICRA. IEEE International Conference on Robotics and Automation, vol. 1, IEEE, 2001, pp. 109–114, https://doi.org/10.1109/robot.2001.932538.
[41] J. Park, M. Chang, I. Jung, H. Lee, K. Cho, 3D printing in the design and fabrication of anthropomorphic hands: a review, Advanced Intelligent Systems 6 (2024 Apr), https://doi.org/10.1002/aisy.202300607.
[42] S. Roccella, M.C. Carrozza, G. Cappiello, J.J. Cabibihan, C. Laschi, P. Dario, H. Takanobu, M. Matsumoto, H. Miwa, K. Itoh, A. Takanishi, Design and development of five-fingered

hands for a humanoid emotion expression robot, International Journal of Humanoid Robotics 04 (2007 Mar) 181–206, https://doi.org/10.1142/s0219843607000996.
[43] A. Georgopoulou, L.M. Eckey, F. Clemens, Soft chemiresistive sensing shields soft robotic actuators from mechanical degradation due to critical solvent exposure, Advanced Engineering Materials 26 (2024 Apr), https://doi.org/10.1002/adem.202301723.
[44] L.U. Odhner, L.P. Jentoft, M.R. Claffee, N. Corson, Y. Tenzer, R.R. Ma, M. Buehler, R. Kohout, R.D. Howe, A.M. Dollar, A compliant, underactuated hand for robust manipulation, The International Journal of Robotics Research 33 (2014 Feb) 736–752, https://doi.org/10.1177/0278364913514466.
[45] J.J. Cabibihan, M.K. Abubasha, N. Thakor, A method for 3-D printing patient-specific prosthetic arms with high accuracy shape and size, IEEE Access 6 (2018) 25029–25039, https://doi.org/10.1109/access.2018.2825224.
[46] D.A. Bennett, S.A. Dalley, D. Truex, M. Goldfarb, A multigrasp hand prosthesis for providing precision and conformal grasps, IEEE/ASME Transactions on Mechatronics 20 (2015 Aug) 1697–1704, https://doi.org/10.1109/tmech.2014.2349855.
[47] T. Do, T. Tjahjowidodo, M. Lau, S. Phee, An investigation of friction-based tendon sheath model appropriate for control purposes, Mechanical Systems and Signal Processing 42 (2014 Jan) 97–114, https://doi.org/10.1016/j.ymssp.2013.08.014.
[48] S. Jacobsen, E. Iversen, D. Knutti, R. Johnson, K. Biggers, Design of the Utah/M.I.T. Dextrous Hand, in: Proceedings. 1986 IEEE International Conference on Robotics and Automation, IEEE, 1986, https://doi.org/10.1109/robot.1986.1087395.
[49] J.E. Huber, N.A. Fleck, M.F. Ashby, The selection of mechanical actuators based on performance indices, Proceedings of the Royal Society of London. Series A: Mathematical, Physical and Engineering Sciences 453 (1997) 2185–2205, https://doi.org/10.1098/rspa.1997.0117.
[50] L. Birglen, T. Lalibert, C.M. Gosselin, Underactuated Robotic Hands, Springer Publishing Company, Incorporated, 2008.
[51] H. Liu, D. Yang, S. Fan, H. Cai, On the development of intrinsically-actuated, multisensory dexterous robotic hands, ROBOMECH Journal 3 (2016 Feb), https://doi.org/10.1186/s40648-016-0043-5.

Chapter 13

Artificial finger and hand fabrication

13.1 Introduction

From the designs of robotic and prosthetic hands over the years, we can observe that many of them can be traced back to classical designs that made use of machined parts with their geometries reflecting a combination of 3-dimensional primitive shapes consisting of cylinders, half-spheres, and boxes with rounded or chamfered edges. While they communicate a deterministic purpose to be functional and mechanical, the preference of prosthesis users is for their hands to not draw unnecessary attention. For humanoid robots, artificial hands that look soft and humanlike can communicate safety, companionship, and being more human-friendly.

Artificial hands address a psychosocial purpose for an amputee. Prosthetic hands help ward off the social stigma associated with the loss of hands [1]. Instead of concealing use of prosthetic hands, the traditional designs of prosthetic hands attract unwanted attention [2][3][4]. It has been suggested that the general aspects of the human hand need to be replicated in order to create a natural look for prosthetic hands. The ideal features include lifelike shape, size, color, and fine details, such as pores and hairs [5][6][7].

Additive manufacturing technologies extend the stock cylindrical and metal block parts for machining, enabling us to design more organic shapes to replicate the surfaces and contours of the human body. In addition, advanced imaging techniques, such as computed tomography (CT) and 3D scanning [8], allow us to extract the digital versions of the human body for further processing into the parts we need for reconstruction. Soft materials are then used in the molds that have been constructed from the digital models.

The next sections are organized according to the workflows used to construct artificial fingers (Sec. 13.2) and hands (Sec. 13.3). Each of those sections describes the participant's data and how they were collected. Then, the construction process is described. This process consists of the digital modeling and digital fabrication techniques used. The subsections end with the photographs of the constructed prototypes and the relevant measurements. Sec. 13.4 summarizes the outcomes of this chapter.

Humanizing Artificial Touch. https://doi.org/10.1016/B978-0-32-395519-5.00025-8

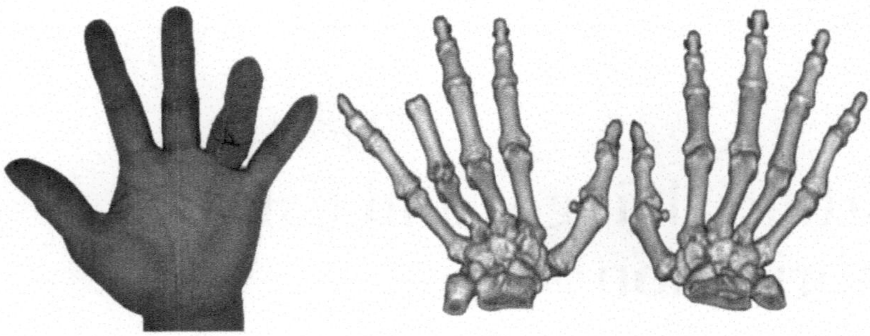

(a) (b) (c)

FIGURE 13.1 The patient's injured finger and the computed tomography (CT) scan. (a) The injured left ring finger with a hand-sewn covering. The CT images of the bone structure of the patient's (b) left and (c) right hands. The extent of amputation is shown at the ring finger of the left hand. Reproduced from Cabibihan [6], ©2011 Cabibihan. This is an open-access article distributed under the terms of the Creative Commons Attribution License (CC BY 3.0).

13.2 Finger fabrication

13.2.1 Participant

The participant of this case study was a middle-aged female individual, then aged 48, who experienced the loss of her fourth digit, specifically the ring finger, on her non-dominant hand, specifically the left hand. This occurred as a result of an accident involving a laundry spinner. The surgical procedure involved the removal of the affected part of the finger at the proximal interphalangeal joint. The participant has applied a bandage-like device to conceal her missing digit (Fig. 13.1a). The individual expressed a desire for a passive finger prosthesis with the aim of attaining a visually satisfactory aesthetic of her fingers. Approval for the experimental protocol used in this case study was granted by the Institutional Review Board of the National University of Singapore. The participant granted written informed consent for the publication of the CT images and photographs.

13.2.2 Construction process

13.2.2.1 Scanning process

The design process consisted of two parts. First, the patient-specific data were acquired through a CT scanner. Second, the data were processed through Computer-Aided Design (CAD) methodology. The patient's data were acquired utilizing a helical CT scanner (Philips Brilliance, Koninklijke Philips Electronics, Netherlands). The scanner was employed to acquire images of both the injured and unaffected areas of the hand.

The parameters used in the scan were as follows:

- voltage of 140 kV
- current of 240 mA

- gantry tilt angle of 0°
- table advance rate of 2 mm/s
- reconstruction interval of 2 mm

Figs. 13.1b and 13.1c illustrate the digitally reconstructed images of the patient's left and right hands. The CT images were stored using the Digital Imaging and Communications in Medicine (DICOM) format. The DICOM files were imported into the Mimics software (Materialise, Belgium) in order to visualize the three-dimensional geometry of the patient's data. The design and meshing operations were conducted using the 3-Matic software (Materialise, Belgium).

13.2.2.2 Design

In Cabibihan [6], the design was executed in the following manner.

1. A mirror image operation was conducted on the skin tissue of the unaffected finger.
2. The bone was replicated using a mirroring process. Both the skin tissue and bone geometries were located in the stump region. Fig. 13.2a

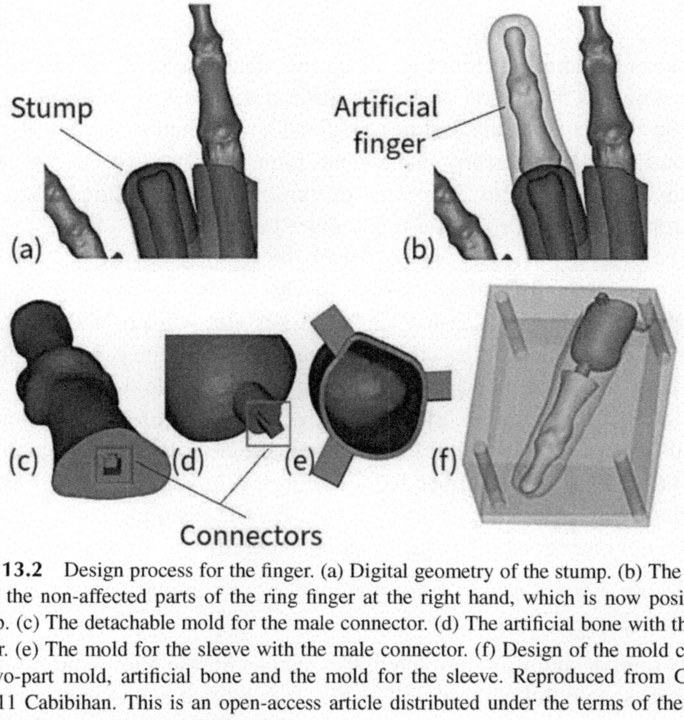

FIGURE 13.2 Design process for the finger. (a) Digital geometry of the stump. (b) The mirrored image of the non-affected parts of the ring finger at the right hand, which is now positioned at the stump. (c) The detachable mold for the male connector. (d) The artificial bone with the female connector. (e) The mold for the sleeve with the male connector. (f) Design of the mold consisting of the two-part mold, artificial bone and the mold for the sleeve. Reproduced from Cabibihan [6], ©2011 Cabibihan. This is an open-access article distributed under the terms of the Creative Commons Attribution License (CC BY 3.0).

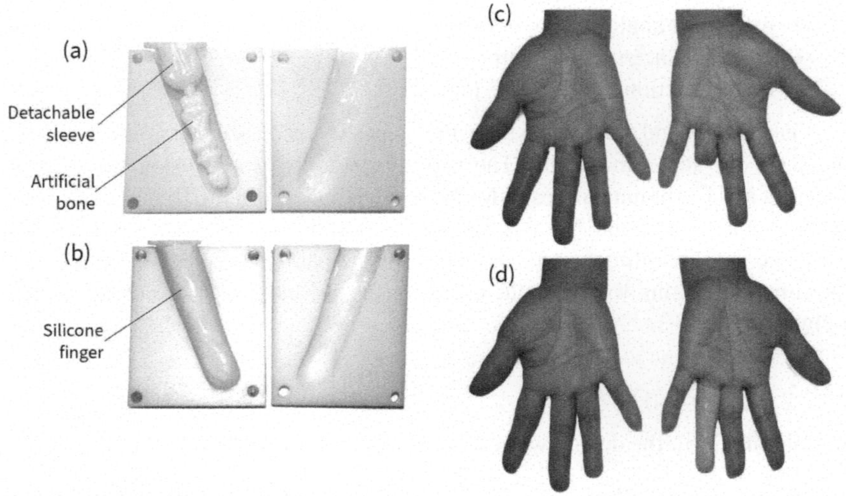

FIGURE 13.3 Prosthetic finger on the mold and on the amputee. (a) The molds of the patient's reconstructed bone, which was accurately positioned using the fins. (b) The prosthetic finger after the curing process. (c) The amputated ring finger before the fitting of the prosthetic finger. (d) The prosthetic finger after fitting. No color matching techniques were implemented on the sample shown. Reproduced from Cabibihan [6], ©2011 Cabibihan. This is an open-access article distributed under the terms of the Creative Commons Attribution License (CC BY 3.0).

shows the stump, while Fig. 13.2b illustrates the placement of mirrored geometries of the non-affected bone and skin tissue on the stump.

3. The geometry of the stump was considered when extending the sleeve towards the metacarpophalangeal joint. The sleeve had a material thickness of 1.5 mm. A proper design of a sleeve ensures a secure and proper fit of the prosthetic finger onto the stump.
4. The sleeve was made to be detachable from the artificial bone. This allows reusability of the molds in case additional fingers need to be fabricated (Figs. 13.2c and 13.2d). A key slot was provided to facilitate the identification of the proper orientation during the assembly process.
5. Three fins were added to the openings of the mold to ensure precise alignment of both the bone and sleeve mold structures in relation to the silicone skin. These fins effectively constrained all the necessary degrees of freedom of the embedded structures (Fig. 13.2e). A two-part split mold was used for casting (Fig. 13.2f). The resulting cavity represents the volume occupied by the silicone material.

13.2.2.3 Fabrication

After the mold design was completed, the designs for the base, bone, and sleeve were saved as stereolithography (STL) files for additive manufacturing process. The part was made using a 3D printer (Model 350, Objet Geometries Ltd., USA) and Fullcure VeroWhite resin (Code 830, Objet Geometries Ltd., USA). The machine prints at 600 dpi on the x and y axes and 1600 dpi on the z axis. The accuracy is within 0.1 mm.

Silicone (GLS40, Prochima, s.n.c., Italy) was injected into the mold with a syringe. For casting with silicone, bubbles could form, which will result in poor appearance on the skin surface. A vacuum chamber was used to remove bubbles that formed during the pouring process. Full curing can be achieved after 24 hours.

13.2.3 Results

Figs. 13.3a and 13.3b display the molds prior to and following the curing process of silicone. The silicone finger was extracted from the base mold. The sleeve's mold was then removed from the bone to obtain the final product. The prosthetic finger was subsequently delivered to the recipient.

Figs. 13.3c and 13.3d display the photographs taken prior to and following the fitting process. The finger prosthesis was secured to the finger by utilizing the vacuum effect created by the stump. A dress ring serves the purpose of enhancing aesthetics and concealing the junction. The patient expressed contentment regarding the size, form, and proper fit of the prosthetic finger.

13.3 Hand fabrication

13.3.1 Participant

A 26-year-old female patient underwent an amputation. An immediate below-elbow amputation of the left arm was performed following the accident. The length of the residual limb was determined to be approximately 93 mm, as measured from the crease of the elbow joint.

The methodology employed in this study did not involve any invasive or possibly dangerous techniques and adhered to the ethical guidelines outlined in the Code of Ethics of the World Medical Association, including the Declaration of Helsinki. The patient provided written informed consent for the publication and dissemination of her CT scan data and photographs.

13.3.2 Construction process

13.3.2.1 Scanning process

In order to gather data for the purpose of designing a prosthetic arm, computed tomography (CT) images were obtained from both the affected and non-affected arms of the patient. The patient's data were obtained using a helical CT scanner (Somatom Definition AS, Siemens, Germany).

The CT images were stored in the Digital Imaging and Communications in Medicine (DICOM) format. The DICOM files were imported into Mimics Innovation Suite (Materialise, Belgium) to facilitate image processing and enable visualization of the patient's data in three-dimensional (3D) geometry. Figs. 13.4a and 13.4b show the CT scan images of the affected and unaffected arms, respectively.

The patient's scan parameters were as follows:

- voltage of 130 kV
- current of 70 mA
- gantry tilt angle of 0°
- table advance rate of 1 mm/s
- reconstruction interval of 2 mm

In order to ensure the accuracy and reliability of the findings, the CT scan images of the constructed prosthetic arm were obtained (Fig. 13.4c). These images were utilized to conduct a comparative analysis between the participant's original arm and the replicated arm. The prosthetic arm under investigation was subjected to scanning using a helical CT scanner (Somatom Drive, Siemens, Germany).

The parameters used in the scan of the prosthetic arm were as follows:

- voltage of 120 kV
- current of 60 mA
- gantry tilt angle of 0°
- table advance rate of 1 mm/s
- reconstruction interval of 2 mm

13.3.2.2 Design

The CT data of the patient were imported into a 3D modeling software (3-Matic, Materialise, Belgium). This software was utilized for the purpose of visualizing

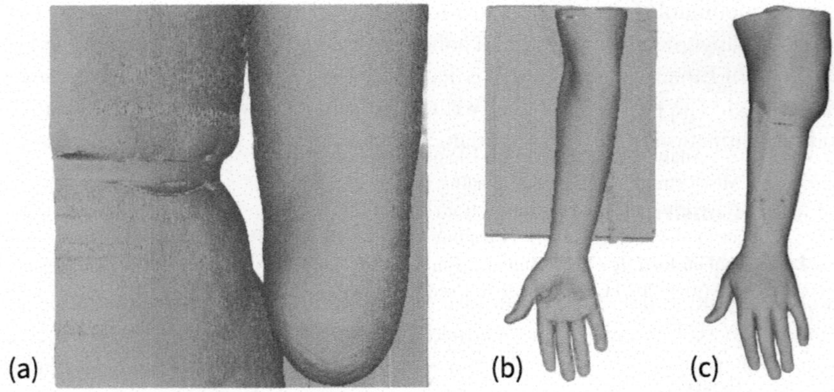

FIGURE 13.4 Computed Tomography (CT) images. (a) Amputated left arm. (b) Uninjured right arm. (c) Prosthetic arm. Reproduced from Cabibihan et al. [7]. ©2018 Cabibihan et al. This is an open access article distributed under the terms of the Creative Commons Attribution License (CC BY 4.0).

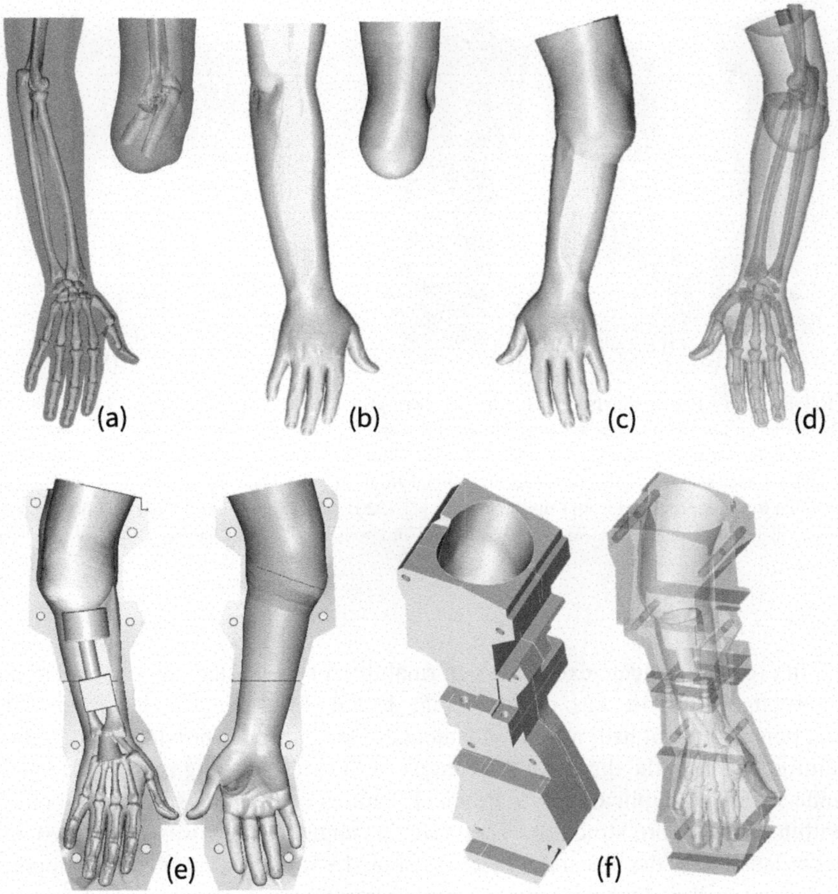

FIGURE 13.5 Design process for the prosthetic arm. (a) The patient's arms, one affected and one non-affected. (b) Three-dimensional surface reconstruction of the computed tomography (CT) scan images. (c) The right arm is positioned in a mirror image of the left arm and reconstructed digitally. (d) The transparent view of the reconstructed arm showing the digital replication of the elbow joint, ulna, and radial bones, which were used for alignment purposes. (e) The reconstruction of the missing arm with simplified skeleton for support. (f) The four-part model of the mold. The completed prototype of the mold design showing the cavity where the liquid silicone material is poured. Reproduced from Cabibihan et al. [7]. ©2018 Cabibihan et al. This is an open access article distributed under the terms of the Creative Commons Attribution License (CC BY 4.0).

and editing the data, as depicted in Figs. 13.5a and 13.5b. The software was employed to generate a symmetrical representation of the right arm for the left arm, as depicted in Fig. 13.5c. In order to achieve precise anatomical alignment of the absent left forearm, the radius and ulna bones were employed as fiducial landmarks for the mirrored left forearm (Fig. 13.5d).

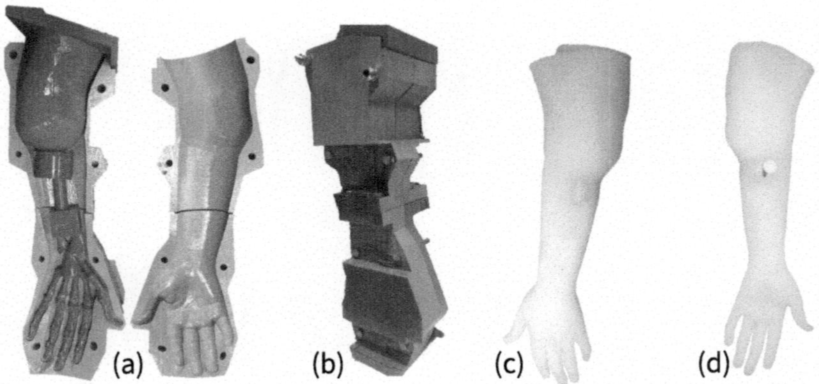

FIGURE 13.6 Fabrication process for the prosthetic arm. (a) The 3D printed outputs for the mold. Once the liquid silicone has fully solidified, the cup-shaped component (shown as red part) is removed in order to position the socket. (b) The mold is secured in place using bolts and nuts. (c) The dorsal surface of the constructed artificial limb. (d) The volar surface of the constructed artificial limb. Reproduced from Cabibihan et al. [7].

In order to mitigate excessive deformability of the casted silicone material, a structural support was devised (Fig. 13.5e). Furthermore, this particular component functioned as a fundamental element that provided structural reinforcement to the digits. In the context of the present design, it was deemed unnecessary to duplicate the anatomical features of the radius and ulna bones within the support structure. Therefore, a singular shaft was developed in accordance with the research objectives. Fig. 13.5f shows the skeleton support and the overall mold that is suitable for four-part printing. The size was designed to be accommodated by the build volume of a desktop 3D printer (Maker-Bot Industries, NY, USA; dimensions: $29.5 \times 19.5 \times 16.5\ mm^3$).

13.3.2.3 Fabrication

Fig. 13.6a shows the 3D printed skeletal structure of the hand and arm, as well as the four-part split mold. The assembled mold and skeleton structure are shown in Fig. 13.6b, where they were securely fastened using bolts and nuts for fixation purposes. The present design investigates the utilization of a mold as a means to create a cavity for the deposition of liquid silicone material (Dragonskin, Smooth-On Inc, PA, USA). At ambient conditions, the silicone's curing process took 24 hours. The prosthetic arm is shown in Figs. 13.6c and 13.6d. The prototype was completed within a span of five days. In this study, we will henceforth designate the reflected counterpart of the uninjured limb as the patient's arm.

13.3.3 Results

In order to assess the accuracy of the fabricated prosthetic arm and socket, a comparative analysis was conducted between their respective computed tomography (CT) scan images and those obtained from the amputee. The present study necessitated the alignment or superimposition of the models. The geometric matching of the digital models was performed using the Global Registration feature of the design software (3-Matic, Materialise, Belgium). In this study, the computational model was constructed by filling up the spaces occupied by the bone and internal structures, as the calculations solely focused on comparing the external topography.

Three distinct similarity measures were employed to assess the degree of resemblance between the variables under investigation. In this work, the Dice Similarity Coefficient (DSC) [9] was used as a primary metric for evaluating the similarity between two entities. The measure under consideration is formally defined as twice the intersection of the parts that are being compared. The intersection of the two-part models can be obtained by performing a Boolean intersection operation. The DSC was determined as:

$$DSC = \frac{2|V_1 \cap V_2|}{|V_1| + |V_2|} \tag{13.1}$$

where V_1 is the geometric model of the part coming from the amputee's CT data, and V_2 is the geometric model from the prosthetic arm. The present study investigates the relationship between the value of the DSC and the volume of two geometries. Specifically, the aim was to determine whether the DSC approaches a value of 1 when the volumes of the two geometries become more similar.

The second similarity metric employed in this work was the percentage error, denoted as the difference between the mirrored computed tomography (CT) scan data obtained from the unaffected arm of the amputee and the corresponding data acquired from the prosthetic arm, expressed as a percentage relative to the CT scan data derived from the unaffected arm. The presence of surplus material can be observed in this context. The percentage error was determined as follows:

$$Error = \frac{|V_2 - V_1|}{|V_1|} \times 100 = \frac{|V_3|}{|V_1|} \times 100 \tag{13.2}$$

where $V_2 - V_1$ denotes the Boolean subtraction (i.e., the geometric region present in the prosthetic model V_2 that does not overlap with the amputee's model V_1), rather than the scalar algebraic difference between their total volumes. V_3 is the geometric model of this surplus material originating from the prosthetic arm, and $|V_3|$ is its volumetric measurement.

A Boolean subtraction operation was employed to compare the volume of the uninjured arm with the volume of the prosthetic arm. The presence of a

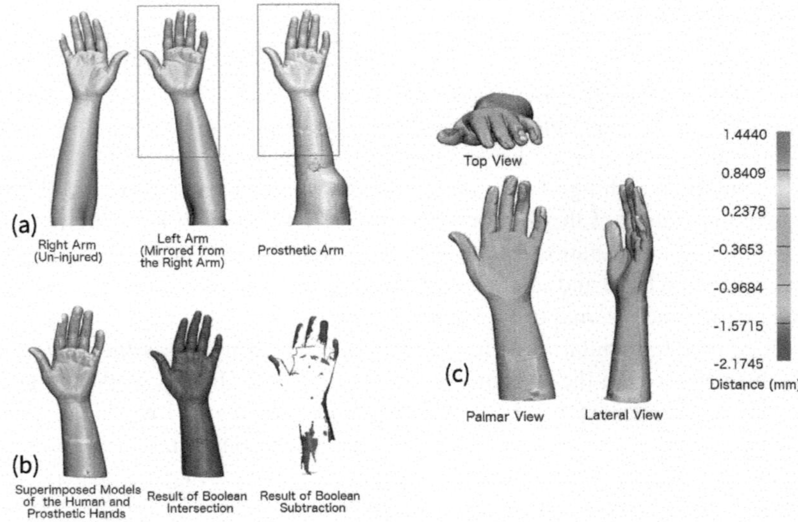

FIGURE 13.7 Comparisons between the prosthesis and the patient's mirrored arm image. (a) The unaffected arm of the patient and its mirror image, along with the CT scan information of the prosthetic arm. The region of interest for comparisons is shown in red. (b) The human and artificial hands overlaid on top of one another. Also shown are the results of performing Boolean intersection and subtraction procedures. (c) Comparative analysis showing the different views of the digitally-constructed arm. The color bar represents the maximum distance between the prosthetic arm model and the human arm model, which is indicated by the color red and measures 1.44 mm. The smallest distance, indicated by the color blue, measures 2.17 mm. Reproduced from Cabibihan et al. [7]. ©2018 Cabibihan et al. This is an open access article distributed under the terms of the Creative Commons Attribution License (CC BY 4.0).

minimal percentage error suggests a high degree of resemblance between the two digital data under examination.

The final similarity measure employed in this study was based on distance measurements. A comparative analysis of parts was conducted utilizing 3-Matic software (Materialise, Leuven, Belgium). The software computes the point-by-point descriptive statistics pertaining to measures of central tendency, specifically the mean and median, as well as measures of dispersion, including the standard deviation, minimum, and maximum values.

In accordance with the principles of anatomical symmetry, the prosthetic arm was developed to replicate the mirrored geometry of the non-injured arm. In order to assess the accuracy of replicating the unaffected arm, the specific region of interest (ROI) was determined as shown in Fig. 13.7a.

The patient's arm model has a volume of 813.79 cm^3, whereas the prosthetic arm demonstrated a slightly higher volume of 817.19 cm^3. These findings shed light on the quantitative differences in volume between the two arm models, providing valuable insights for the development and optimization of prosthetic limb designs. In order to determine the DSC, both models were overlaid for comparison (Fig. 13.7b). The present study employed the Boolean

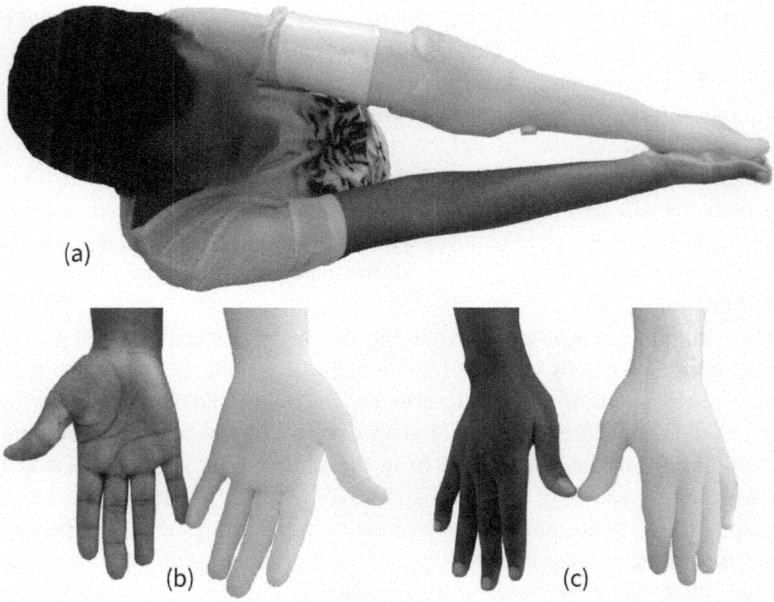

FIGURE 13.8 Comparisons between the non-affected arm and the prosthesis. (a) The participant comparing the length of the prosthetic arm to her unaffected arm. Shape and length comparisons (b) at the volar and (c) dorsal sides of the hand. Reproduced from Cabibihan et al. [7]. ©2018 Cabibihan et al. This is an open access article distributed under the terms of the Creative Commons Attribution License (CC BY 4.0).

intersection operation to ascertain the shared volume between the two models. The experimental procedure yielded a volume measurement of 779.91 cm^3. The method proposed in this study demonstrated a remarkable DSC value of 0.96. This finding indicates a strong agreement between the prosthetic arm and the mirrored representation of the patient's injured arm.

In order to determine the origins of the errors, a Boolean subtraction operation was employed to conduct an inquiry into the surplus material existing between the prosthetic arm and the CT model of the patient's unaffected arm, which had been mirrored. The Boolean subtraction operation was performed, resulting in a difference of 5.46 cm^3. The excess material, denoted as a percentage error, was found to be 0.67%. The excess materials were predominantly localized at the distal regions of the fingers, as depicted in Fig. 13.7c. In Fig. 13.8a the patient compares the lengths of the artificial arm and the non-affected arm. The hands' volar sides (Fig. 13.8b) and dorsal sides (Fig. 13.8c) were also compared for shape and size.

13.4 Summary

Understanding the deeper meanings and requirements associated with the appearance of artificial hands motivates us to find and/or develop technologies and workflows to address these underlying needs, such as the psychosocial role attached to prosthetic hands. This chapter presented a streamlined workflow for the fabrication of prosthetic fingers and hands. Advancing beyond the machining technologies used 30 years ago, technologies such as imaging, computer-aided design, and additive manufacturing technologies were explored. In the examples shown, the technique made use of computed tomography (CT) data from both the affected and unaffected limbs, allowing for the reconstruction and mirroring of anatomical features with precision. Having digital models and fabrication techniques at their disposal, designers can minimize the trial-and-error approach.

The methodology was validated through two case studies. The first was for a prosthetic finger for a 48-year-old female with a domestic accident. The second was for a below-elbow prosthetic arm for a 26-year-old female following a car accident. For both cases, the CT data were requested from both amputees. The uninjured limb and the stump from the injured part were requested. The data were processed with a specialized CAD software. Mirroring techniques were applied to ensure anatomical symmetry and a more natural appearance. Connectors, attachments, fluid flow paths for the casting with silicone material were designed using the same software. The molds were 3D printed and the usual procedures for material casting and curing are used.

To test the results, the fabricated hand was again subjected to a CT scan. Having the digital model of the uninjured hand and the fabricated hand, the models can easily be digitally overlapped. One model can be subtracted from the other. With this method, the Dice Similarity Coefficient (DSC) between the prosthetic and the uninjured arm was 0.96 out of 1.0, which indicates near-perfect overlap in terms of volume. The percent error in volume was 0.67%. The mean distance between the surfaces that were compared was −0.34 mm (SD = 0.48 mm). The maximum and minimum differences between the fabricated hand and the uninjured hand were 1.44 mm and −2.17 mm, respectively.

Knowing the implied or hidden meanings of why a prosthetic hand or a robot hand is needed can guide the development of new technologies or combine available technologies to digitally and physically construct the hand structures to address fully or partially the design requirements [6][7][10]. The scanning-CAD modeling-3D printing-casting-measurement approach described in this chapter offers a closed-loop approach in humanizing the construction of the structures representing human arms, hands or fingers.

References

[1] C.D. Murray, The social meanings of prosthesis use, Journal of Health Psychology 10 (2005 May) 425–441, https://doi.org/10.1177/1359105305051431.

[2] M. Pilley, D.N. Quinton, Digital prostheses for single finger amputations, The Journal of Hand Surgery: British & European 24 (1999) 539–541, https://doi.org/10.1054/jhsb.1999.0198.

[3] A. Alison, S.E. Mackinnon, Evaluation of digital prostheses, The Journal of Hand Surgery 17 (1992) 923–926, https://doi.org/10.1016/0363-5023(92)90471-z.

[4] A.Y. Kini, P.P. Byakod, G.S. Angadi, U. Pai, A.J. Bhandari, Comprehensive prosthetic rehabilitation of a patient with partial finger amputations using silicone biomaterial: a technical note, Prosthetics and Orthotics International 34 (2010) 488–494, https://doi.org/10.3109/03093646.2010.486391.

[5] E.L. Leow, B.P. Pereira, A.K. Kour, R.W.H. Pho, Lifelikeness in multilayered digital prostheses, Prosthetics and Orthotics International 21 (1997 Apr) 40–51, https://doi.org/10.3109/03093649709164529.

[6] J.J. Cabibihan, Patient-specific prosthetic fingers by remote collaboration–a case study, PLoS ONE 6 (2011) 1–6, https://doi.org/10.1371/journal.pone.0019508.

[7] J.J. Cabibihan, M.K. Abubasha, N. Thakor, A method for 3-D printing patient-specific prosthetic arms with high accuracy shape and size, IEEE Access 6 (2018) 25029–25039, https://doi.org/10.1109/access.2018.2825224.

[8] A. Gaballa, L.A. Lambert, K. Diab, J.J. Cabibihan, Image processing of 3D scans for upper limb prosthesis of the war-wounded, in: IEEE 20th International Conference on Bioinformatics and Bioengineering (BIBE), 2020, pp. 596–601, https://doi.org/10.1109/bibe50027.2020.00102.

[9] L.R. Dice, Measures of the amount of ecologic association between species, Ecology 26 (1945) 297–302, https://doi.org/10.2307/1932409.

[10] J.J. Cabibihan, M. Mudassir, Parametrically-Designed Artificial Hand with Multifunctional Grasps. Social Robotics, Springer Nature, Singapore, 2025, pp. 81–92, https://doi.org/10.1007/978-981-96-1151-5_9.

Chapter 14

Artificial flexible finger joints

14.1 Introduction

The dexterity of the human hands can be attributed to its structure. Each hand has 27 bones and over 20 joints [1][2]. There is also a complex arrangement of ligaments, cartilage, and tendons. Muscles generate the gross or fine movements to stabilize the grasps and to modulate the grip forces. Named from the palm to the fingertip, the joints of the fingers: the metacarpophalangeal (MCP), proximal interphalangeal (PIP) and distal interphalangeal (DIP) joints are crucial for the hand's movements. These joints function to help the human hand achieve adduction/abduction of the five fingers with the radial adduction/abduction of the thumb, the flexion/extension of the five fingers with adduction/abduction of the palm, and the retroposition/opposition of the thumb with the bending/flattening of the palm (see Fig. 14.1; [3]).

Replicating the human joints with artificial ones has been one of the design tasks needed to be accomplished in artificial hand design. Early designs utilized hinged mechanisms that were constructed from steel or aluminum materials. Such designs offered good properties for strength, but had limitations on compliance. The recent designs have taken advantage of additive manufacturing and soft robotics technologies to replicate some positive aspects of the mechanical behavior and ease of manufacture [4].

This chapter presents the design and analysis techniques to investigate flexible artificial finger joints. The focus of the analysis was to achieve the robustness of the joints so that they can achieve the range of motion (ROM) similar to human joints as well as to withstand hyperextension. The Finite Element Analysis (FEA) approach was used to evaluate the performance of three designs and two material types under 5 N load. Based on the optimal results, the flexible joint design was fabricated and was used as joints for the fingers of an artificial hand. It was tested to compare with the basic pose of human hands in their neutral position.

The next sections were organized according to the structures for human and artificial finger joints. In Sec. 14.2, the human finger joints are discussed in relation to the anatomical structure, biomechanics of motion, and the range of motion. Sec. 14.3 discusses the design and materials aspects of artificial finger joints. Sec. 14.4 presents an example of the design and analysis of flexible joints, followed by Sec. 14.5 where a case study of the flexible joints was applied to replicate the natural hand pose. Sec. 14.6 summarizes the chapter.

Humanizing Artificial Touch. https://doi.org/10.1016/B978-0-32-395519-5.00026-X

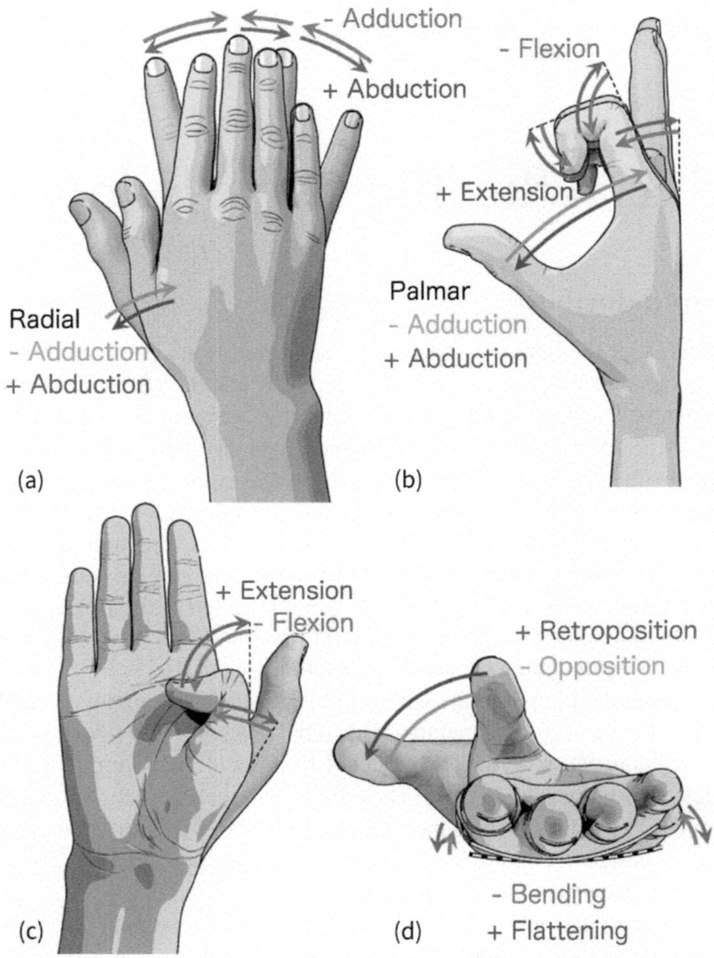

FIGURE 14.1 Various movements of the fingers. (a) The radial abduction/adduction of the thumb and the relative abduction/adduction of the remaining fingers. (b) The palmar abduction/adduction of the thumb and the flexion/extension of the remaining fingers. (c) The flexion/extension of the remaining thumb joints. (d) The opposition/retroposition of the thumb and the bending/flattening of the palm. The blue/purple and red/green arrows indicate the positive and negative directions, respectively. Reproduced from Cabibihan et al. [3]. ©2020 Cabibihan et al. Used under Creative Commons CC BY 4.0 license.

14.2 Human finger joints

14.2.1 Anatomical structure of the finger joints

The fingers of the human hand are dexterous due to the synchronized coordination from the actuation from the muscles to the joints. Except for the thumb, there are three joints that connect the three phalanges of the fingers (i.e., proximal, middle, and distal phalanges). These joints are the metacarpophalangeal (MCP), proximal interphalangeal (PIP) joint, and the distal interphalangeal (DIP) joint (see Fig. 14.2; [1]). The MCP joint allows flexion/extension, abduction/adduction, and limited rotation. The PIP and DIP joints are hinge joints that permit flexion/extension [5]. The thumb has two joints, the interphalangeal (IP) hinge joint and MCP joint, which allow opposition. To minimize friction, the surfaces of the bones are covered with cartilage. The extensor mechanism, which is the arrangement of tendons and ligaments, has its significant role in the extension movement of the fingers [2].

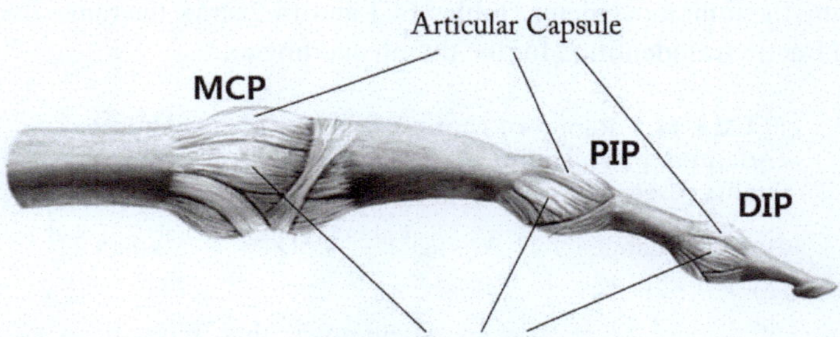

FIGURE 14.2 Phalanges of the human finger with the MCP, PIP, and DIP joints. Each joint has an articular capsule and a pair of collateral ligaments. Reproduced from Lee and Choi. [6]. ©2020 Lee and Choi. Used under Creative Commons CC BY 4.0 license.

14.2.2 Biomechanics of finger joint motion

When the hand grips and releases an object, the movement of the finger joints involves the coordinated motion of flexion and extension of the MCP, PIP, and DIP joints. In descending order, the joint with the largest range of motion (ROM) is the PIP, followed by MCP, and the DIP. For fast hand grasps, the motion starts from the PIP flexion before the DIP and the MCP. Among the three joints, the PIP has the highest peak velocities. The finger joints move through the coordinated movements of the extrinsic and intrinsic muscles. The extrinsic muscles, including the flexor digitorum profundus (FDP) and flexor digitorum superficialis (FDS), have their origins in the forearm and extend to the fingers through tendons [7]. The intrinsic muscles, composed of the lumbricals and interossei, facilitate the fine motor control of the fingers and the stability of the grasp [8].

The intrinsic finger muscles also play a crucial role in finger joint motion and stability. Valero-Cuevas et al. [9] demonstrated that large fingertip forces can be generated by subject-independent patterns of intrinsic muscle activation, highlighting their importance in dexterous manipulation. Kursa et al. [10] found that the forces in the intrinsic muscles are not dependent on the rate of fingertip loading during an isometric task, suggesting their role in stabilizing the finger joints.

14.2.3 Range of motion

The range of motion (ROM) of the joints was obtained in activities of daily living [11]. The activities included: holding a phone, holding a can, using a zipper, holding a toothbrush, turning a key, using a comb, writing with a pen, holding a fork, holding scissors, unscrewing a jar, and holding a hammer. Standard measurement methods together with an electrogoniometer were used to measure the joint movements. Tables 14.1 and 14.2 show the range of motion for the 11 activities identified for the thumb and fingers.

TABLE 14.1 Range of Motion in ADLs of the Thumb. Adapted from Hume et al. [11].

Joint	Range (°)	Average (°)	SD (°)	Median (°)
MCP	10–32	21	±5	22
IP	2–43	18	±5	19

TABLE 14.2 Range of Motion in ADLs of the Fingers. Adapted from Hume et al. [11].

Joint	Range (°)	Average (°)	SD (°)	Median (°)
MCP	33–73	61	±12	62
PIP	36–86	60	±12	63
DIP	20–61	39	±14	39

Hume et al. [11] also reported the normal ROM for the thumb and the rest of the fingers. For the MCP, the ROM was 0–56° and for IP it was −5° to 73°. For the ROM of the four other fingers, the MCP has a range of 0–100°, the PIP 0–105°, and the DIP 0–85°. The ROM for the ADLs has a lower range of values as compared to the normal ROM of the finger joints.

14.3 Artificial finger joints

14.3.1 Artificial joints design

Among the possible mechanisms for the joints of the artificial fingers, the revolute joint is the simplest type to achieve the rotational motion of the finger joints. Although commonly used, revolute joints are limited in their ability to replicate the coupled movements and compliance of human fingers.

In Xu et al. [12], they proposed artificial fingers that were constructed from 3D printed finger bones, silicone rubber sleeves, and crocheted ligaments. Their design achieved joint motions of 30–90° for the MCP, 0–110° for the PIP, and 0–70° for the DIP. They intended this configuration to replicate the biomechanic and dynamic properties of the human finger.

14.3.2 Materials

The material to be used for the joints has a significant effect on the performance characteristics, weight, durability, and ease of fabrication. The materials typically used were those compatible with the subtractive manufacturing processes (i.e., machining). These were typically steel and aluminum. In contrast, additive manufacturing processes, particularly with 3D printing, have been used more recently due to its ease of use and availability. The materials that are compatible with such processes are polymeric materials, which are typically PLA, ABS, and TPU. These normally come as filaments that are commercially available for 3D printing.

With 3D printed molds, the other typically used materials for joints are soft, elastomeric materials (e.g., silicones and polyurethanes). Soft materials such as these provide the advantage of compliance and impact absorption. They help improve robustness and add the possibility to tolerate overextensions to prevent the failure of the joints.

14.4 Example: design and analysis of flexible joints

In Alkhatib et al. [13], we described the design and analysis of a flexible artificial joint that can be robust against hyperextension of the joints. The joints are constructed from elastomeric materials, offering superior flexibility compared to conventional plastic joints. The design looks into the critical need for durability and minimal maintenance, addressing the concerns for repair and maintenance.

14.4.1 Methodology

14.4.1.1 Representative sample of an artificial hand

The Flexy Hand was used as a representative example of openly accessible, 3D-printed prosthetic hands (Fig. 14.3a). The Flexy Hand is a body-powered prosthesis controlled by wrist movement, employing non-elastic cables for flexion to enable grasping or holding objects. The flexion mechanism operates by transferring the motion generated by the wrist's bending to the cables attached near the fingertips. To achieve hand extension and release objects, the design utilizes flexible joints that enable structural elastic deflection for opening the hand. In contrast, other openly accessible 3D printed prosthetic hand designs, such as the Cyborg Beast Hand [14], Phoenix Hand [15], and Raptor Reloaded Hand [15][16], rely on elastic cables as the primary mechanism for the extension of joints.

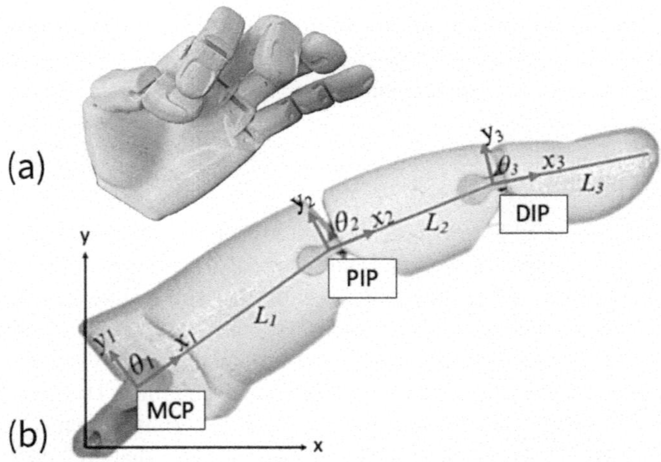

FIGURE 14.3 A representative design of a 3D printed artificial hand. (a) A typical design of an artificial hand with passive joints. (b) An artificial finger with soft and flexible joints showing the links, joints, and frame assignments. Reproduced from Alkhatib et al. [13], with permission. ©2019 IEEE. All rights reserved.

14.4.1.2 Finger kinematics

To compute the final position of the fingertip, forward kinematics analysis was performed, employing the Denavit-Hartenberg (D-H) parameters. The D-H methodology [17] provides a systematic approach for constructing transformation matrices based on four parameters derived from the link frame of a robotic system (as detailed in Table 14.3). The joint angles θ_1, θ_2, and θ_3 correspond to the metacarpophalangeal (MCP), proximal interphalangeal (PIP), and distal interphalangeal (DIP) joint angles, respectively. Fig. 14.3b illustrates the variable assignments associated with the links and joints of the finger.

TABLE 14.3 D-H Parameters of the 3D Printed Prosthetic Finger.

Link	a_i	α_i	d_i	θ_i
1	L_1	0	0	θ_1
2	L_2	0	0	θ_2
3	L_3	0	0	θ_3

The X and Y positions of the fingertip can be calculated from Eq. (14.1) and (14.2) as:

$$X = L_1 \cos\theta_1 + L_2 \cos(\theta_1 + \theta_2) + L_3 \cos(\theta_1 + \theta_2 + \theta_3) \tag{14.1}$$

$$Y = L_1 \sin\theta_1 + L_2 \sin(\theta_1 + \theta_2) + L_3 \sin(\theta_1 + \theta_2 + \theta_3) \tag{14.2}$$

where the lengths of the proximal, middle, and distal phalanges correspond to $L_1 = 44$ mm, $L_2 = 29$ mm, and $L_3 = 27$ mm, respectively. The MCP, PIP, and DIP joint angles are denoted as θ_1, θ_2, and θ_3, respectively.

14.4.1.3 Numerical simulations

The artificial hand functions through its flexible joints, with each joint exhibiting one degree of freedom. These joints are crucial to the design because of the continuous bending they experience. A numerical analysis was conducted to assess the performance of these joints. Two fundamental design elements were examined: the joint geometry and the material employed. In Alkhatib et al. [13], three geometries were examined (Fig. 14.4). The simulation results of the three geometries are shown in Fig. 14.5 for comparison. The two materials evaluated were polyurethane (Polytek, Polytek Development Corp, USA) and silicone (Dragon Skin, Smooth-On, USA). The mechanical characteristics of these materials are specified in Table 14.4.

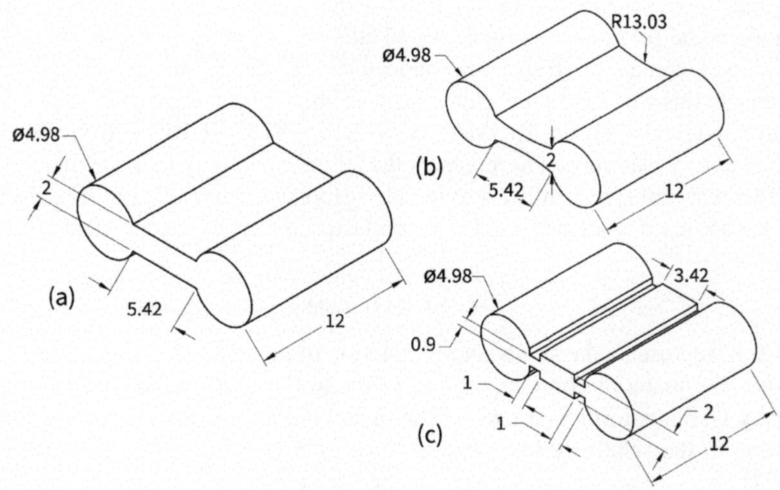

FIGURE 14.4 Joint designs for the numerical studies: (a) design 1, (b) design 2, and (c) design 3.

TABLE 14.4 Mechanical Properties of the Materials Used.

Material	Mass Density (g/cm^3)	Ult. Tensile Strength (MPa)	Poisson's Ratio
Polyurethane	1.01	1.50	0.49
Silicone	1.07	3.30	0.45

The non-linear analysis was conducted using a Finite Element Analysis software (LS-DYNA, Livermore Software Technology Corporation, USA). Nonlinear modeling is advised for deformable materials. Silicone and polyurethane

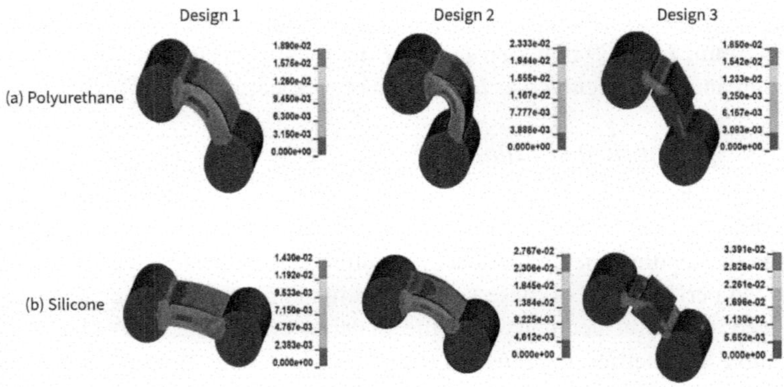

FIGURE 14.5 Simulation results on the stress concentration areas of the three designs for (a) polyurethane material and (b) silicone material.

exhibit nonlinear behavior beyond their elastic limits [18][19][20]. The flexible joints were partitioned into small 3D hexahedral elements with 8 nodes and a single nodal rotation to mitigate the likelihood of rotational distortion. Optimization of element mesh size was conducted based on the internal energy of the designs. In this work, 0.1 mm^3 elements were chosen for meshing all the designs in our numerical analysis. The MAT HYPERELASTIC RUBBER from the LS-DYNA library was chosen to represent the silicone and polyurethane materials.

This material type conforms to the Neo-Hookean model [21] in forecasting the behavior of incompressible hyperelastic materials under deformation (Eq. (14.3))

$$W = C_{10}(I_1 - 3) \tag{14.3}$$

where W represents the strain energy function of the hyperelastic material, C_{10} denotes the material constant, and I_1 refers to the first invariant of the right Cauchy-Green deformation tensor. The model did not incorporate damage theory due to the relatively low applied force of 5 N, as noted by Alhaddad et al. [22].

14.4.1.4 Fabrication of the artificial hand and joints

The open-sourced 3D printed artificial hand was fabricated using PLA filament. The printing process required approximately 17 hours. The polyurethane material was utilized for casting within the 3D printed molds. The joints required 24 hours to cure. Subsequent to fabrication, these joints were inserted into the slots between the segments of the fingers.

14.4.2 Results

14.4.2.1 Simulation results

The flexible joint designs were subjected to a load at one end with a 5 N distributed load. The findings were obtained after 0.35 seconds. Fig. 14.5 shows that the stress concentration area for the flexible joint was situated at the fixed end of the design. The peak stresses recorded were 0.189 MPa for polyurethane and 0.143 MPa for silicone. The polyurethane exhibited greater flexibility than the silicone. The *Y*-displacement value for polyurethane was 2.49 mm while the measured value for silicone was 0.975 mm.

From the simulation results, the polyurethane material showed higher flexibility as compared to the silicone material, when a force of 5 N was applied in the simulations. From the ROM of the human finger joints, the artificial finger joints need to show ROM angles from 0° to 100°. As compared to silicone, the polyurethane material is a more suitable choice as it exhibits the necessary flexibility to achieve the required ROM.

14.4.2.2 Finger flexion results

To verify the kinematic analysis of the middle finger and assess the functionality of the flexible joints, the middle finger was exposed to various bending angles, and the final *X* and *Y* coordinates of the fingertip were documented (Fig. 14.6; [13]). Five joint angles were utilized: 0°, 20°, 40°, 60°, and 80°. In all instances, the angles were established as equal (i.e., $\theta_1 = \theta_2 = \theta_3$). The experimental observations were compared with the theoretical values derived from the kinematic analysis (Eqs. (14.1) and (14.2)). Table 14.5 displays the theoretical and experimental end positions of the middle finger in the *X* and *Y* coordinates.

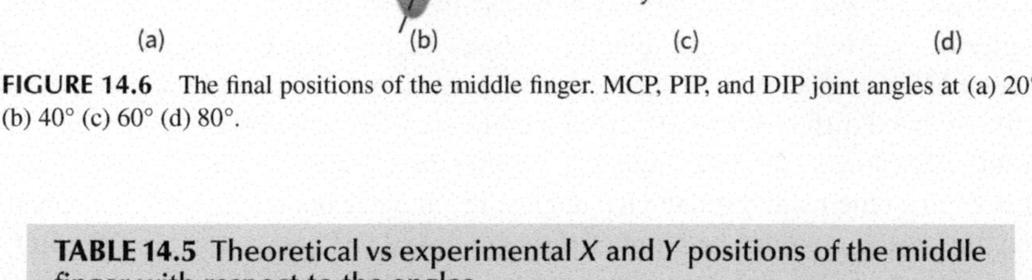

FIGURE 14.6 The final positions of the middle finger. MCP, PIP, and DIP joint angles at (a) 20° (b) 40° (c) 60° (d) 80°.

TABLE 14.5 Theoretical vs experimental *X* and *Y* positions of the middle finger with respect to the angles.

Angles	Theoretical		Experimental		Error	
	X (mm)	*Y* (mm)	*X* (mm)	*Y* (mm)	*X* (%)	*Y* (%)
0°	100.0	0.0	100.0	0.0	0.0	0.0
20°	77.1	57.1	78.0	59.0	1.2	3.4
40°	25.3	80.2	24.0	81.0	5.1	1.0
60°	−19.5	63.3	−19.0	64.0	2.3	1.2
80°	−33.1	29.9	−33.0	30.0	0.3	0.2

14.5 Case study: natural hand pose with flexible joints

Typical robotic or prosthetic hands are oriented in an open palm posture. This is not a natural posture for humans. When a person is in a standing position, his or her hands are posed in a limp posture and not simply a flat and stiff position. Flexible joints were found to be ideal in replicating the natural pose of human hands.

14.5.1 Methodology

Four volunteers were recruited in an experimental measurement of their hands to collect the data. The data gathering employed no harmful or unsafe methods and adhered to the World Medical Association's Code of Ethics (Declaration of Helsinki). The participants have executed a consent form and consented to provide images of their arms for research purposes, in the condition of anonymization of their identities.

The participants were instructed to stand upright and a snapshot of their hands was captured. The measurements were conducted with a four-step procedure.

1. The natural pose of each participant's hand was photographed.
2. The hands' joint angles were obtained via imaging techniques.
3. The mean angles for the MCP, PIP, and DIP joints were calculated.
4. The 3D printed artificial hand was calibrated to achieve the natural posture derived from the experiment participants.

14.5.2 Results

The joint angles of the index fingers of both the human and artificial hands are shown in Fig. 14.7. In more detail, the joint angle data of the human hand and the artificial hand were compared in Table 14.6. Results demonstrate that the angles of the MCP of the artificial hand closely approximate the mean values of human MCP angles, with differences not exceeding $\pm 1^\circ$. The PIP angles of the artificial hand differed by $\pm 10^\circ$ from the human PIP angles. With the exception of the index finger, the DIP angles of the prosthetic hand exhibited a difference of $\pm 2^\circ$ from the mean human DIP angles. From these data, it can be concluded that the natural pose can be replicated for MCP and DIP joints. For the PIP joint, the difference was larger than the expected angles due to the non-elastic cables that are not directly connected to the PIP joint.

14.6 Summary

Understanding the joints of the human fingers, its anatomical structure, biomechanics, and the range of motion give us the possibility to design and analyze artificial joints that can replicate the performance of the human fingers for robotic or prosthetic hands. For grasping and manipulation tasks, the fingers

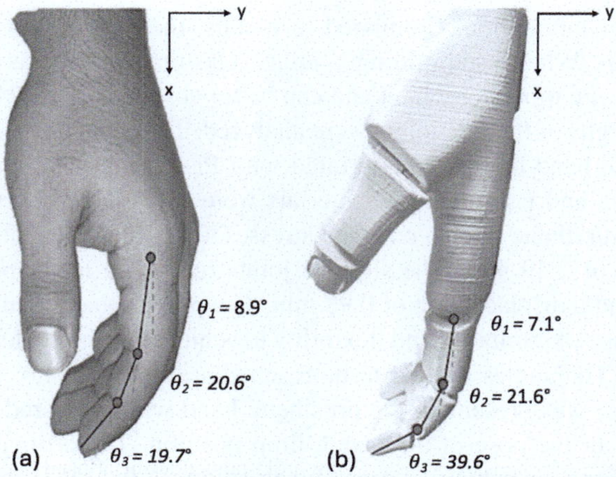

FIGURE 14.7 Joint angles in the natural pose for (a) the human hand, and (b) 3D printed artificial hand. Reproduced from Alkhatib et al. [13], with permission.

TABLE 14.6 Comparison of Human and Artificial Fingers.

Finger	MCP Angle (°)		PIP Angle (°)		DIP Angle (°)	
	Human	Artif.	Human	Artif.	Human	Artif.
Thumb	25.2±5.8	23.8	–	–	16.9±6.2	18.2
Index	7.20±1.4	7.1	34.8±10.0	21.6	16.6±3.1	39.6
Middle	17.2±3.8	16.2	31.4±5.2	22.0	21.9±2.0	21.2
Ring	19.4±3.6	18.0	37.8±7.2	20.9	16.8±7.5	16.8
Little	11.8±3.4	12.0	24.3±6.5	11.3	20.9±10.0	19.2

flex or extend depending on the task. The human system has pain sensors and a reflex system that activates when joints overextend their limits. Artificial joints do not have such mechanisms. They could just fail when they reach their limits [23].

The chapter first described the human finger joints as the inspiration source. The human hand's dexterity is due in part to the metacarpophalangeal (MCP), proximal interphalangeal (PIP) and distal interphalangeal (DIP) joints. These work together with the cartilage, ligaments, and synovial fluid [12]. The extrinsic muscles (i.e., flexor digitorum profundus and superficialis) and the intrinsic muscles (i.e., interossei and lumbricals) work in coordination to achieve synchronized motions and obtain grasp stability [7].

The joints on robotic or prosthetic hands were typically constructed with metallic materials with revolute mechanisms. The metallic joints were beneficial for their strength and manufacturability. Their main drawbacks include the lack of compliance, which would have an effect on limited movements. More recent works have used additive manufacturing with the standard 3D printing filaments, such as PLA and TPU [4]. These polymeric materials are lightweight and easier to be used for fabrication. Compared to metals, the 3D printed parts exhibit lower strength. When used as finger joints, elastomers, such as silicone and polyurethane, are more compliant and can better absorb impact [24].

Three designs of flexible joints were analyzed. The joints were implemented in a prosthetic hand that was 3D printed with PLA materials. For the flexible joints, silicone and polyurethane materials were investigated. Results showed that the polyurethane joints reached stresses of 0.189 MPa and a vertical displacement of 2.49 mm. The silicone joints reached a lower peak stress of 0.145 MPa and a displacement of 0.98 mm. The polyurethane joints provide an optimal balance of durability and flexibility to achieve the finger flexions needed for grasping. The angles needed to be reached are in the 0°–80° range. In the case study that was presented, the prosthetic hand was compared to the hands of volunteers in the pose of a relaxed, limp position. The artificial hand with the flexible joints can achieve comparable positions of the MCP and DIP angles within ±2° of human joints. The PIP joint angle had a difference of up to 10°.

Knowing the basic information of the joints of the human fingers provides guidance on how to replicate this for artificial fingers. The kinematic modeling and the FEA calculations were able to show us the range of motion that artificial fingers were able to reach during the flexion needed to grasp objects. The flexible joints have an added advantage of providing additional compliance when the fingers overextend, thus increasing robustness, which can increase the life of the artificial fingers.

References

[1] J.M.F. Landsmeer, Anatomical and functional investigations on the articulation of the human fingers, Acta Anatomica Supplementum 25 (1955) 1–69.

[2] W.P. Cooney, E.Y. Chao, Biomechanical analysis of static forces in the thumb during hand function, The Journal of Bone and Joint Surgery 59 (1977) 27–36.

[3] J.J. Cabibihan, F. Alkhatib, M. Mudassir, L.A. Lambert, O.S. Al-Kwifi, K. Diab, E. Mahdi, Suitability of the openly accessible 3D printed prosthetic hands for war-wounded children, Frontiers in Robotics and AI Volume (2021) 7–2020, https://doi.org/10.3389/frobt.2020.594196.

[4] D. Denizon, S. Dogru, L. Marques, Optimization of an underactuated two finger robotic hand using genetic algorithms, in: Walking Robots into Real World, Springer Nature, Switzerland, 2024, pp. 28–40, https://doi.org/10.1007/978-3-031-71301-9_3.

[5] N. Fowler, A. Nicol, Interphalangeal joint and tendon forces: normal model and biomechanical consequences of surgical reconstruction, Journal of Biomechanics 33 (2000 Sep) 1055–1062, https://doi.org/10.1016/s0021-9290(00)00080-4.

[6] G. Lee, Y. Choi, Bio-inspired tendon-driven finger design with isomorphic ligamentous joint, IEEE Access 8 (2020) 18240–18251, https://doi.org/10.1109/access.2020.2968739.

[7] A.D. Nimbarte, R. Kaz, Z.M. Li, Finger joint motion generated by individual extrinsic muscles: a cadaveric study, Journal of Orthopaedic Surgery and Research 3 (2008 Jul), https://doi.org/10.1186/1749-799x-3-27.

[8] J. Diong, M.E. Héroux, S.C. Gandevia, R.D. Herbert, Minimal force transmission between human thumb and index finger muscles under passive conditions, PLoS ONE 14 (2019) e0212496, https://doi.org/10.1371/journal.pone.0212496.

[9] F.J. Valero-Cuevas, F.E. Zajac, C.G. Burgar, Large index-fingertip forces are produced by subject-independent patterns of muscle excitation, Journal of Biomechanics 31 (1998 Aug) 693–703, https://doi.org/10.1016/s0021-9290(98)00082-7.

[10] K. Kursa, E. Diao, L. Lattanza, D. Rempel, In vivo forces generated by finger flexor muscles do not depend on the rate of fingertip loading during an isometric task, Journal of Biomechanics 38 (2005 Nov) 2288–2293, https://doi.org/10.1016/j.jbiomech.2004.07.035.

[11] M.C. Hume, H. Gellman, H. McKellop, R.H. Brumfield, Functional range of motion of the joints of the hand, The Journal of Hand Surgery 15 (1990) 240–243, https://doi.org/10.1016/0363-5023(90)90102-w.

[12] Z. Xu, V. Kumar, Y. Matsuoka, E. Todorov, Design of an Anthropomorphic Robotic Finger System with Biomimetic Artificial Joints, 2012 4th IEEE RAS & EMBS International Conference on Biomedical Robotics and Biomechatronics (BioRob), IEEE, 2012, pp. 568–574.

[13] F. Alkhatib, E. Mahdi, J.J. Cabibihan, Design and analysis of flexible joints for a robust 3D printed prosthetic hand, in: 2019 IEEE 16th International Conference on Rehabilitation Robotics (ICORR), 2019, pp. 784–789.

[14] J. Zuniga, D. Katsavelis, J. Peck, J. Stollberg, M. Petrykowski, A. Carson, C. Fernandez, Cyborg beast: a low-cost 3d-printed prosthetic hand for children with upper-limb differences, BMC Research Notes 8 (2015) 10.

[15] Enabling the future. Cyborg Beast Hand (2015), http://enablingthefuture.org/current-design-files/cyborg-beast-hand/. (Accessed 21 October 2025).

[16] F. Alkhatib, J.J. Cabibihan, E. Mahdi, Data for benchmarking low-cost, 3D printed prosthetic hands, Data in Brief 25 (2019) 104163, https://doi.org/10.1016/j.dib.2019.104163.

[17] P. Corke, Robotics Vision and Control: Fundamental Algorithms in MATLAB, vol. 118, Springer, 2017.

[18] J.J. Cabibihan, S. Pattofatto, M. Jomaa, A. Benallal, M. Carrozza, P. Dario, The conformance test for robotic/prosthetic fingertip skins, in: The First IEEE/RAS-EMBS International Conference on Biomedical Robotics and Biomechatronics. BioRob 2006, 2006, pp. 561–566, https://doi.org/10.1109/biorob.2006.1639148.

[19] J.J. Cabibihan, M.C. Carrozza, P. Dario, S. Pattofatto, M. Jomaa, A. Benallal, The uncanny valley and the search for human skin-like materials for a prosthetic fingertip, in: 2006 6th IEEE-RAS International Conference on Humanoid Robots, 2006, pp. 474–477.

[20] J.J. Cabibihan, S. Pattofatto, M. Jomaa, A. Benallal, M.C. Carrozza, Towards humanlike social touch for sociable robotics and prosthetics: comparisons on the compliance, conformance and hysteresis of synthetic and human fingertip skins, International Journal of Social Robotics 1 (2009) 29–40, https://doi.org/10.1007/s12369-008-0008-9.

[21] R.W. Ogden, Non-Linear Elastic Deformations. Dover Civil and Mechanical Engineering, Dover Publications, Mineola, NY, 1997.

[22] A.Y. Alhaddad, S.E. AlKhatib, R.A. Khan, S.M. Ismail, A.S.S. Shehadeh, A.M. Sadeq, J.J. Cabibihan, Toward 3D printed prosthetic hands that can satisfy psychosocial needs: grasping force comparisons between a prosthetic hand and human hands, in: A. Kheddar, E. Yoshida, S.S. Ge, K. Suzuki, J.J. Cabibihan, F. Eyssel, He H. Cham (Eds.), Social Robotics, Springer International Publishing, 2017, pp. 304–313, https://doi.org/10.1007/978-3-319-70022-9_30.

[23] A. Al-Adbah, J.J. Cabibihan, Assessing the Mobility of a Prosthetic Hand Featuring Soft Silicone Joints for War-Wounded Children: A Case Study, 2025 47th Annual International Conference of the IEEE Engineering in Medicine and Biology Society (EMBC), IEEE, 2025, pp. 1–4, https://doi.org/10.1109/embc58623.2025.11253345.

[24] T.S. Castaneda, P. Capsi-Morales, X. Zhang, Piazza C. Bio-Inspired, Spherical Joints for a Soft Articulated Artificial Hand. 2024 10th IEEE RAS/EMBS International Conference for Biomedical Robotics and Biomechatronics (BioRob), IEEE, 2024, pp. 1307–1312, https://doi.org/10.1109/biorob60516.2024.10719752.

Chapter 15

Artificial underactuated hands

15.1 Introduction

From the daily activity of eating with utensils to complex highly dexterous suturing or knot-tying in surgery, the human hand exhibits gross to fine control of the digits to achieve simple to difficult grasping and manipulation tasks. Around 39 musculotendon units actuate the movement of 27 bones in the hand, combining to orchestrate movement with over 25 degrees of freedom (DOF) [1]. Forces are transmitted to the fingers through the tendons that are connected to the extrinsic muscles and are located in the forearm. Within the hand are intrinsic muscles that enable fine postures and stable grasps [2]. To adjust grip forces and prevent objects from slipping, fingertip sensors and joint proprioceptors work in tandem to provide the motor cortex with real-time information about the object [3][4].

This chapter elaborates more on the actuation mechanisms of the human and artificial hands. The next section presents the actuation of the human hands from the perspective of their anatomy and the activation patterns and control. The succeeding section compares the differences between grasping and manipulation and the implication of each one in their design requirements. For simple grasping tasks, underactuated mechanisms (i.e., there are fewer actuators than the number of DOFs) offer an optimized solution for the design of robot hands, where the design uses fewer motors and incorporates linkages, springs, and cables to compensate for missing actuators (e.g., [5][6][7]).

The chapter demonstrates the workings of an underactuated mechanism in the analysis of a representative 3D printed artificial hand [8][9]. Kinematic modeling was conducted to quantify the fingertip trajectory as a function of joint angles. Analysis on the cable tension was also conducted to determine the cable forces to the limits of the joint flexion. With this sample of a 3D printed hand, it was demonstrated that only three grasp types can be achieved out of the 33 human grasp types.

Sec. 15.2 briefly presents the actuation mechanism of the human hand, where the anatomy, activation patterns, and control are described. Sec. 15.3 discusses the actuation of artificial hands, where grasping and manipulation are compared and the mechanism of underactuation was defined. Sec. 15.4 describes the methodologies to analyze the kinematics and forces involved during the grasp of a 3D printed underactuated hand. Sec. 15.5 supplements the previous section to

Humanizing Artificial Touch. https://doi.org/10.1016/B978-0-32-395519-5.00027-1

analyze the material and the cable used to assemble the hand. Finally, Sec. 15.6 summarizes the findings in the chapter.

15.2 Actuation of human hands

15.2.1 Anatomy of the musculoskeletal structure of the finger

The movement of the human hand is made possible by the 39 extrinsic and intrinsic musculotendon units in the human hand, which control the movement of 27 bones [10]. In coordination, that mechanism is able to perform movements of over 25 degrees of freedom (DOF). The extrinsic muscles in the forearm are long flexor and extensor muscles that impart force to the fingers through tendons. The intrinsic hand muscles are involved in sophisticated motor control and dexterity of the hand (Fig. 15.1a; [11]). The action of the interosseous and lumbrical muscles is shown in Fig. 15.1b to produce flexion of the MCP and extension of the PIP joint.

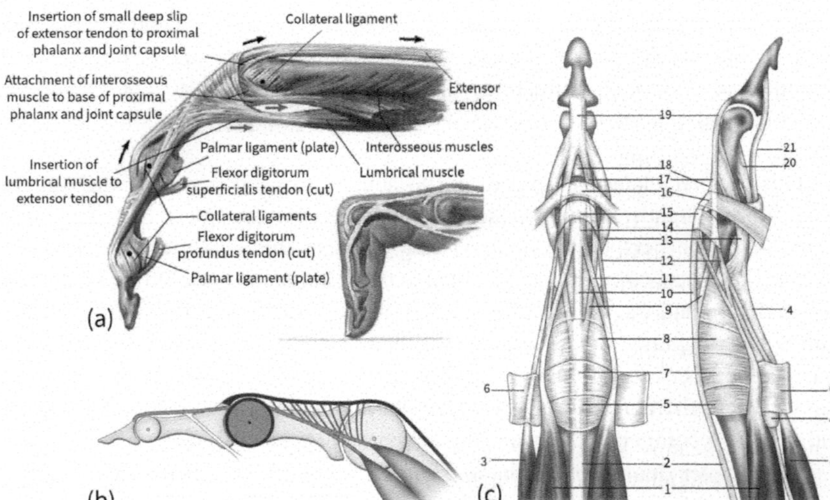

FIGURE 15.1 Anatomy of the finger structure. (a) Intrinsic muscles of a finger contribute to the extensor apparatus (mechanism) of the finger. (b) Schematic drawing of extensor apparatus showing the action of the interosseous and lumbrical muscles in producing flexion of the MCP and extension of the PIP joint. (c) The extensor apparatus of the finger. (1) Interosseous muscle. (2) Extensor communis tendon. (3) Lumbrical muscle. (4) Flexor tendon fibrous sheath. (5) Sagittal bands. (6) Intermetacarpal ligament. (7) Transverse fibers of extensor apparatus. (8) Oblique fibers of the extensor apparatus. (9) Lateral band of extensor tendon. (10) Central or middle band/slip. (11) Central or middle band of interosseous tendon. (12) Lateral band of interosseous tendon. (13) Oblique retinacular ligament (Landsmeer's ligament). (14) Middle extensor tendon. (15) Spiral fibers. (16) Transverse retinacular ligament. (17) Lateral extensor tendons. (18) Triangular ligament. (19) Terminal extensor tendon. (20) Flexor superficialis tendon. (21) Flexor profundus tendon. Reproduced from Schreuders et al. [11], with permission.

Tendons are used to transmit force from the muscles to the bones of the fingers [1]. The flexor digitorum superficialis (FDS) and flexor digitorum profundus (FDP) flex the fingers, and the extensor digitorum communis (EDC) extends the fingers [10]. These tendons are arranged anatomically and include anatomical pulleys that increase the mechanical advantage by keeping the tendons near the bones, thus reducing the energy spent in the movement of the tendons [12]. Shown in Fig. 15.1c is the detailed extensor apparatus of the finger.

Lee and Kamper [13] reported that the four DOFs of the index finger (e.g., the MCP, PIP, and DIP joints) are driven by seven muscles (i.e., EDC, EIP, FDS, and FDP) and three intrinsic muscles (i.e., FDI, FPI, LUM). They added that even to produce an isometric fingertip force requires all seven muscles. Such a mechanism implies the idea of muscle synergies rather than one-muscle/one-joint pattern to accomplish a task.

15.2.2 Activation patterns and control

In Lee and Kamper [13], they showed that the muscle activation patterns are determined by the hand posture, the tendons, and the passive mechanics. Based on their intramuscular recording studies, the muscle activation patterns significantly vary with interphalangeal (IP) posture. The MCP posture had little effect on muscle activation, although it had a larger effect on muscle length. Moreover, they reported that the extensor hood collects tendons from 4–5 muscles and redistributes their forces to the finger phalanges; fingertip forces can be produced without changing the activation patterns.

The hood's mechanics let the CNS reuse one synergy across configurations. For the motor commands, the CNS selects the muscle activation patterns that minimize the overall muscle effort when subjected to biomechanical constraints. As described in Fig. 1.2, the control of the actuation of the fingers makes use of the feedforward and feedback mechanisms, where corrective actions are triggered by the tactile and proprioceptive sensors [3].

15.3 Actuation of artificial hands

15.3.1 Comparisons: grasping vs manipulation

Birglen et al. [14] compared the definition of grasping and manipulation. They defined grasping as the property of a system to prevent motion of an object. Manipulation was defined as the capability of a system to control the position and orientation of an object.

Birglen et al. [14] further argued that a device optimized for robust grasping lacks the dexterity for complex manipulation. On the contrary, a highly dexterous hand struggles to achieve stable grasps of unpredictable objects.

To address this dilemma, designers have used passive elements and mechanical intelligence [5][15][16][6]. Such an approach can enhance the adaptation of the fingers, which can then help in keeping and maintaining a stable grasp. As a consequence of this, there will be fewer actuators than DOFs. This implies that complexity can be reduced, but the manipulation capabilities will be limited as compared to fully actuated designs.

15.3.2 Definition

Underactuated mechanisms for artificial hands were defined as mechanisms where the number of actuators is less than the number of degrees of freedom (DoFs) [14]. This means that fewer motors or other actuators are used to control the many joints in the hand. For completeness, the other actuation classes are fully actuated mechanisms and redundantly actuated mechanisms. They are defined as:

$$\begin{aligned} \text{Underactuated mechanisms} &: N_{DOF} > N_{ACT} \\ \text{Fully actuated mechanisms} &: N_{DOF} = N_{ACT} \\ \text{Redundantly actuated mechanisms} &: N_{DOF} < N_{ACT} \end{aligned} \tag{15.1}$$

where N_{DOF} indicates the number of DOFs and N_{ACT} denotes the number of actuators.

In the design of underactuated mechanisms, passive elements, such as springs, linkages, and mechanical stops, are used to compensate for the missing actuation. When implemented on the fingers of a robot hand, the underactuated mechanism adapts to the shape of the grasped object. A characteristic of an underactuated mechanism is that adaptation to an object does not need individual control of each finger joint.

Underactuated mechanisms are more suitable for grasping tasks while fully actuated hands are better suited for manipulation tasks due to their high degree of control, albeit at the cost of increased complexity. Underactuated mechanisms offer advantages such as simpler control, reduced weight, and complexity, resulting in achieving more natural and more energy-efficient grasping movements.

15.4 Analysis on a simple underactuated hand design

In Cabibihan et al. [9], we investigated whether a simple underactuated artificial hand can achieve a range of motion similar to that of a human hand. More specifically, we checked whether the limits imposed by the structural constraints of the artificial hand can affect its range of motion and confirm whether this is similar to a human hand.

15.4.1 Materials

15.4.1.1 Artificial hand sample

Fig. 15.2 shows a sample of an underactuated prosthetic hand that was used for the analysis (Raptor Reloaded, e-NABLE Community, 2014; [17][8]). A desktop 3D printer (Replicator 5th Generation, MakerBot Industries LLC, Brooklyn, NY, USA; build table: 29.5×19.5×16.5 cm^3) was used to 3D print the CAD file using Polylactic Acid (PLA; MakerBot, USA) material. The following configurations were employed:

1. 0.2 mm layers
2. two shells
3. 35% infill
4. 215°C printing temperature
5. cooling fan activated

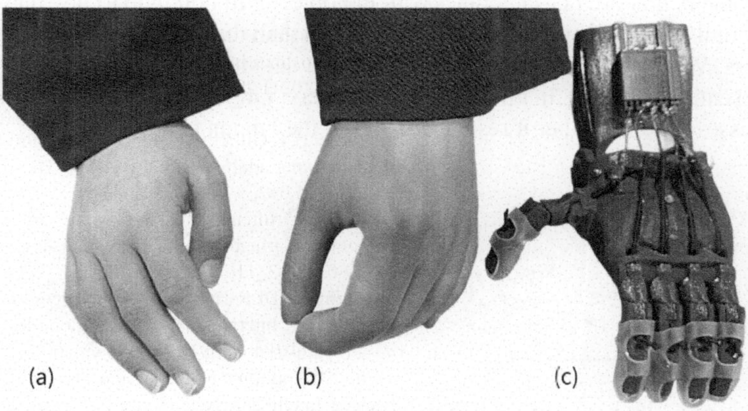

FIGURE 15.2 An amputee with a cosmetic prosthetic hand. (a) The non-injured hand. (b) A worn-out, darkened cosmetic prosthetic hand used by the patient. (c) A sample of an underactuated hand without the cosmetic cover. Reproduced from Cabibihan et al. [9]. ©2020 Cabibihan et al. Used under Creative Commons CC BY 4.0 license.

15.4.1.2 Cables

Both elastic and non-elastic cables were required for the grip and release mechanism in order to finish the assembly. Flexing the fingers requires the use of non-elastic wires (super Dyneema strong braided fishing line, SeaKnight, China). To put the fingers back in their natural position, elastic cords (3 mm dia, Polypropylene Shock Cord, Sgt. Knots Supply Co, NC, USA) were utilized. The cables and cords were securely fastened using a knotting method called the enhanced clinch knot.

15.4.2 Methodology

15.4.2.1 Grasp poses

This sample of artificial hand used in the analyses was a transcarpal prosthetic hand. In order to grasp an object, the user must flex their wrist in order to cause the joints of the artificial hand to flex. The following procedure was followed in order to identify the grasps that this artificial hand can accomplish. First, an 8-year-old child wore the transcarpal prosthetic hand. Second, the child replicated 33 grasps [18] shown in the pictures. The final step was to move the object being grabbed in order to assess the quality of the grasp, following the protocol in [6]. For repeatability, three trials in a row were conducted. The grasp positions that the artificial hand could perform were then shortlisted.

15.4.2.2 Range of motion of the fingers

All of the fingers of an underactuated hand wrap around the object's surface. The fingers that are not in contact with the object will continue to flex until the structural limits are reached if the object is less than the enclosing volume of the fingers. We looked into the boundaries set by the structural constraints for the representative sample. In this section, the fingers' range of motion was evaluated and assessed on whether it resembled that of the human hand.

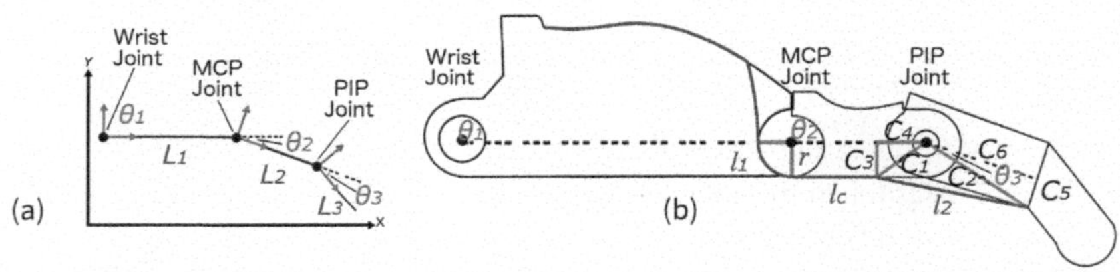

FIGURE 15.3 Illustrative diagrams, meshed finite element model, and materials characterization results. (a) The three links of the index finger showing the three local coordinates and the variables used. (b) The geometrical representation of the finger mechanism showing all the variables to calculate the final position of the fingertip. Reproduced from Cabibihan et al. [9]. ©2020 Cabibihan et al. Used under Creative Commons CC BY 4.0 license.

The *X* and *Y* coordinates of the fingertip were used to establish its locations. The Denavit-Hartenberg (D-H) parameters were determined in order to perform forward kinematics [19]. The index finger's link frame is displayed in Fig. 15.3a. The origin point (0,0) at the wrist joint, where $\theta_1 = 0$, is defined by the two-dimensional Cartesian coordinate system (x, y). The transformation matrices were constructed using the D-H convention and four parameters that were derived from the artificial hand's link frame. These variables include the link lengths a_i, link twists, α_i, link offsets, d_i, and joint angles, θ_i (Table 15.1). The transformation matrices are displayed in Eq. (15.2) to (15.5).

$$
{}^{0}\boldsymbol{T}_1 = \begin{pmatrix} \cos\theta_1 & -\sin\theta_1 & 0 & L_1\cos\theta_1 \\ \sin\theta_1 & \cos\theta_1 & 0 & L_1\sin\theta_1 \\ 0 & 0 & 1 & 0 \\ 0 & 0 & 0 & 1 \end{pmatrix} \tag{15.2}
$$

$$^{1}\boldsymbol{T}_2 = \begin{pmatrix} \cos\theta_2 & -\sin\theta_2 & 0 & L_2\cos\theta_2 \\ \sin\theta_2 & \cos\theta_2 & 0 & L_2\sin\theta_2 \\ 0 & 0 & 1 & 0 \\ 0 & 0 & 0 & 1 \end{pmatrix} \tag{15.3}$$

$$^{2}\boldsymbol{T}_3 = \begin{pmatrix} \cos\theta_3 & -\sin\theta_3 & 0 & L_3\cos\theta_3 \\ \sin\theta_3 & \cos\theta_3 & 0 & L_3\sin\theta_3 \\ 0 & 0 & 1 & 0 \\ 0 & 0 & 0 & 1 \end{pmatrix} \tag{15.4}$$

$$^{0}\boldsymbol{T}_3 = {^{0}\boldsymbol{T}_1}{^{1}\boldsymbol{T}_2}{^{2}\boldsymbol{T}_3} \tag{15.5}$$

TABLE 15.1 Denavit-Hartenberg parameters of the 3D printed artificial index finger. Reproduced from Cabibihan et al. [9]. ©2020 Cabibihan et al. Used under Creative Commons CC BY 4.0 license.

Link Num.	Link Length, a_i	Link Twist, α_i	Link Offset, d_i	Joint Angle, θ_i
1	L_1	0	0	θ_1
2	L_2	0	0	θ_2
3	L_3	0	0	θ_3

The position and orientation quantities, $[X, Y, \phi]^T$, can be used to express the posture of the fingertip. The following forward kinematics equations (Eqs. (15.6) and (15.7)) were used to determine the X and Y positions of the fingertip with respect to the wrist joint angle (θ_1), metacarpophalangeal (MCP) joint angle (θ_2), and the proximal interphalangeal (PIP) joint angle (θ_3). The sum of the joint angles θ_1, θ_2, and θ_3 can be used to indicate the orientation of the finger, ϕ (Eq. (15.8)). The X and Y positions and the orientation were calculated as:

$$X = L_1\cos\theta_1 + L_2\cos(\theta_1+\theta_2) + L_3\cos(\theta_1+\theta_2+\theta_3), \tag{15.6}$$

$$Y = L_1\sin\theta_1 + L_2\sin(\theta_1+\theta_2) + L_3\sin(\theta_1+\theta_2+\theta_3), and \tag{15.7}$$

$$\phi = \theta_1+\theta_2+\theta_3. \tag{15.8}$$

15.4.2.3 Cable tension analysis

To determine the X and Y locations of the fingertip, a mathematical relationship between the position of the fingertip and the applied grasping force was proposed. The links and the variables are shown in Fig. 15.3a. The index finger's structure is displayed in Fig. 15.3b. The cables' tension actuated the finger's grasping flexion, while the elastic cords actuated the finger's returning extension motion.

The MCP joint angle (θ_2) and PIP joint angle (θ_3) changed as a result of the flexion and extension. The DIP and PIP joints were coupled together. The cable length, l_{Cable}, which is in contact with the pulleys, the cable length along the phalanges, and the cable length up to the fixed pin joint where the cable is fastened all affected the value of these angles (Eq. (15.9)). In other words, the length of the wire will decrease as more tension force is applied to it. As a result, the fingers will be able to flex more. At the natural position, the cable's length is at its maximum ($\theta_1 = \theta_2 = \theta_3 = 0°$), and at full tension, it is at its shortest ($\theta_1 = 0°; \theta_2 = \theta_3 = 90°$). It should be noted that in the real design, θ_3 will never be zero. The finger analysis would not be impacted if the minimal value of θ_3 is assumed to be 0 for the sake of simplicity.

Eq. (15.10) to (15.13) can be used to calculate the relationship between the cable's length and the MCP joint angle, θ_2, and PIP joint angle, $\theta_3 = 0°$

$$l_{Cable} = l_1 + l_2 + l_c \tag{15.9}$$

$$l_1 = r\left(\frac{\pi}{2} - \theta_2\right) \tag{15.10}$$

$$l_2 = \sqrt{C_1^2 + C_2^2 - 2C_1C_2\cos(90° - \theta_3)} \tag{15.11}$$

$$C_1 = \sqrt{C_3^2 + C_4^2} \tag{15.12}$$

$$C_2 = \sqrt{C_5^2 + C_6^2} \tag{15.13}$$

where l_c is a constant, while l_1 and l_2 are calculated from Eq. (15.10) and Eq. (15.11), respectively. Eq. (15.12) and (15.13) can be used to obtain C_1 and C_2 from actual measurements. It is possible to measure the constant lengths l_c, C_3, C_4, C_5, C_6 and r directly from the design (Table 15.2).

TABLE 15.2 Design constants of the geometry as measured from the Raptor Reloaded Hand (dimensions in mm). Reproduced from Cabibihan et al. [9].

l_c	C_3	C_4	C_5	C_6	r
15.75	6.00	7.25	6.00	20.00	2.25

15.4.3 Results

15.4.3.1 Grasp poses

Only three grasping postures could be accomplished using the sample of artificial hand that was taken into consideration in this study (Figs. 15.4a – 15.4c). These include an inferior pincer grasp of a golf ball (46.4 g), a lateral grasp of a credit card (10 g), and a palmar pinch of a coin (7.7 g). Out of the 33 total grasps that a human hand can perform (cf. Fig. 12.3), the three grasp types shown in Fig. 15.4 are severely limited. Other artificial hands, including both articulated hands (e.g., [20][21]) and passive parametric designs (e.g., [22]), have demonstrated a much wider range of human grasps, which are close to the 33 grasp poses that the human hands can perform [18].

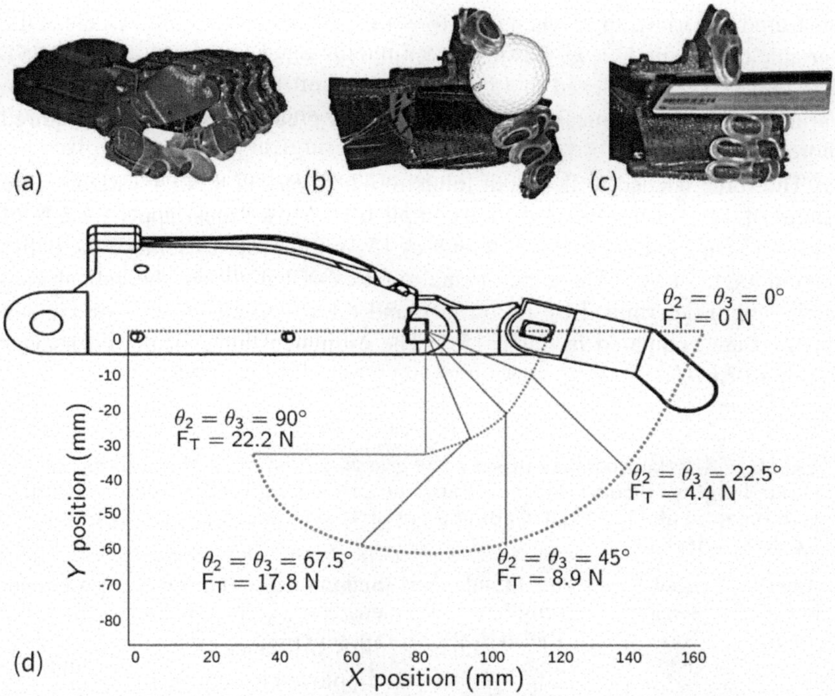

FIGURE 15.4 The limited movements achieved by the representative 3D printed prosthetic hand. (a) Palmar pinch. (b) Inferior pincer grasp. (c) Lateral grasp. (d) The positions of the index finger of the 3D printed prosthetic hand in the X and Y axes with respect to the MCP and PIP joint angles. Reproduced from Cabibihan et al. [9]. ©2020 Cabibihan et al. Used under Creative Commons CC BY 4.0 license.

15.4.3.2 *Range of motion*

The results show how the transmitted force, F_T, and the resulting trajectory of the finger phalanges change in accordance with the changes in the joint angles θ_2 and θ_3.

At the initial position of $\theta_2 = \theta_3 = 0°$, the force transmitted by the linkage is zero ($F_T = 0$ N). The transmitted force increases to 4.4 N at 22.5°, 8.9 N at 45°, and 17.8 N at 67.5°. The maximum force of 22.2 N occurs at $\theta_2 = \theta_3 = 90°$.

The finger's trajectory forms an arc-like path, which shows the relationship between angular variation and positional movement. The geometry of the system implies that there is a dependency of force transmission efficiency on angular configuration.

15.4.3.3 *Cable length and tension*

Fig. 15.4d displays the finger's range of motion with respect to θ_1, θ_2, and θ_3. When the cable produced a force of 4.4 N, the joints θ_2 and θ_3 flexed to 22.5° from the starting conditions. Table 15.3 shows the overall cable length, measured experimentally when $\theta_1 = \theta_2 = \theta_3 = 0°$, was 41.45 mm, whereas the variable cable length, l_{VC}, was just 1.2 mm ($l_{VC} = l_{Cable,experimental} - 41.45$). The joints θ_2 and θ_3 flexed to 45° when a force of 8.9 N was applied, and the variable cable length measured was 5.7 mm. A greater flexion angle required more tensile force.

The force needed to flex both joints to 67.5° was 17.8 N. A variable wire length of 14 mm was needed to accomplish it. A user must apply 22.2 N of force and engage the cable to a length of 15 mm in order to achieve the entire flexion angle of 90°. The experimental total cable length, the theoretical total cable length determined by Eq. (15.9), and the experimental tests previously reported are compared in Table 15.3. The estimated range of the errors was 1.75% to 7.61%.

TABLE 15.3 Relationships between the grasping posture angles, forces, total cable lengths, errors, and the variable cable length. Reproduced from Cabibihan et al. [9]. ©2020 Cabibihan et al. Used under Creative Commons CC BY 4.0 license.

Angle, $\theta_2 = \theta_3$ (deg)	Cable Tension Force (N)	Total Cable Length, Theoretical (mm)	Total Cable Length, Experimental (mm)	Error (%)	Variable Cable Length (mm)
0	0	42.19	41.45	1.75	0
22.5	4.2	37.74	40.45	7.17	1.2
45	8.9	33.22	35.75	7.61	5.7
67.5	17.8	29.34	27.45	6.44	14.0
90	22.2	27.22	26.45	2.83	15.0

15.5 Analysis on the artificial hand's material and cable failure

15.5.1 Finite element modeling

15.5.1.1 Geometry

For the Finite Element Analysis, the geometry of the artificial finger was obtained from the e-NABLE Community and was analyzed in its original geometry. The main contact interaction for this artificial hand occurs at the distal and proximal phalanges. As such, the following parts were modeled in the current analysis: the proximal phalange, the combined intermediate and distal phalanges, the non-elastic cable, the elastic cord, and the pin located at the proximal interphalangeal (PIP) joint (Fig. 15.5a). For computational efficiency, the following were excluded: the wrist, palm, and the pin at the metacarpophalangeal (MCP) joint.

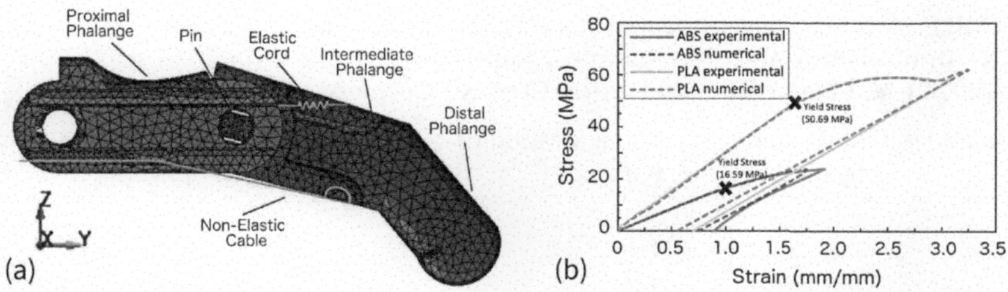

FIGURE 15.5 Model geometry and material stresses. (a) The finite element model of the index finger. The model included two finger phalanges and the pin at the PIP joint. The non-elastic cables and elastic cords were embedded within the structure. (b) The experimental and numerical stress-strain curves of the ABS and PLA filament materials that were obtained from the tensile tests. The X symbol marks the yield stresses for the two materials. Reproduced from Cabibihan et al. [9]. ©2020 Cabibihan et al. Used under Creative Commons CC BY 4.0 license.

15.5.1.2 Mesh design

The geometry of the finger was discretized into small mesh elements (i.e., 3D quadratic tetrahedron solid elements). These elements have four nodes and one nodal rotation to prevent any rotational deformation. Beam elements were used for the non-elastic cable since they cannot be modeled using solid elements as they have constant cross-sectional properties and their length is much greater than their width and hence are susceptible to distributed loads along their length. To represent the spring-like behavior of the elastic cord, a discrete element was used which has one degree of freedom and two nodes. Based on Hooke's law where there is a linear relationship between the applied force and the resulting displacement, a spring constant of 1,000 N/m was applied. An H-refinement test was carried out to verify the model accuracy. This involves enhancing the model by increasing the number of elements through the reduction of the element size. The initial mesh size of the model was between 2.0 and 2.4 mm. The element size ranged from 1.2 to 1.8 mm in the current analysis.

15.5.1.3 Materials

The following materials were used in the analysis: Acrylonitrile Butadiene Styrene (ABS) and Polylactic Acid (PLA) (Table 15.4). Both of these types of materials are commonly used in 3D printing and are compatible with the 3D printer used in the study (MakerBot Industries LLC, NY, USA). From the LS-DYNA material library, the piecewise linear plasticity material model (MAT024) was employed to represent the ABS and PLA materials for the distal finger phalange, the combined intermediate and proximal finger phalanges, and the pin. For the non-elastic cable, a braided fishing line cable (super Dyneema strong braided fishing line, SeaKnight, China) was modeled using the plastic kinematic material model labeled as MAT003 in LS-DYNA library. This was determined to be appropriate for the study due to the cable's assumption of negligible deformation over time.

TABLE 15.4 Mechanical properties obtained experimentally from the tested 3D printed ABS and PLA samples. Reproduced from Cabibihan et al. [9].

Material	Mass Density (g/cm^3)	Young's Modulus (GPa)	Ultimate Tensile Stress (MPa)	Failure Strain (%)
ABS	1.10	1.40	32.00	1.05
PLA	1.30	3.90	54.00	2.20

15.5.1.4 *Materials validation*

The verification process consisted of conducting experimental tensile tests on the ABS and PLA filaments, which were simulated numerically using LS-DYNA software to study the elasto-plastic behavior. The tensile tests were carried out using a universal testing machine (5969 Series, Instron, USA) at a loading rate of 5 mm/min. Elasto-plastic behavior can be defined as materials that achieve their elastic and plastic behaviors after reaching the yield stress of the material. The stress–strain curves from the experimental tests were used to define the effective plastic strain, which forms when the material yields. Results show that there is a good correlation between the experimental and numerical stress–strain curves, further showing that the material models are valid (Fig. 15.5b). From the experimental tensile tests, the yield stresses obtained for ABS and PLA were 16.59 MPa and 50.69 MPa, respectively.

15.5.1.5 *Boundary conditions*

The first boundary condition involved supporting the hole at the proximal phalange in all directions, except for rotation around the x-axis. This ensures stability while allowing movement. The second condition was for the pin to be fully constrained in all directions. This implies that all DOFs for the pin nodes were restricted to prevent any unwanted motion. Lastly, the intermediate phalange was left free to move in any direction. A frictional contact with a friction constant of 0.3 was set between the pin nodes and the intermediate phalange node to maintain the integrity of the model. The same was done between the proximal phalange and the intermediate phalange.

15.5.1.6 *Loading*

Based on the Standards for Accessible Design [23], the standard force value for single-handed grasping, pinching or twisting of objects for human hands is 22.2 N. Therefore, the load that was applied to the ABS finger model was 5 N and 15 N. The PLA finger model was loaded to 5 N, 15 N and 25 N.

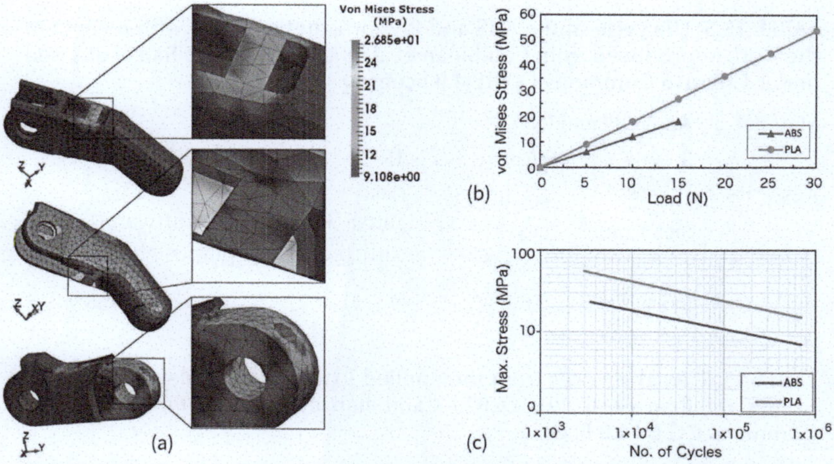

FIGURE 15.6 Analysis results. (a) Finite Element Analysis results showing the locations with the highest von Mises stress under applied loading from the cables being pulled. The maximum von Mises stresses were found at the top and bottom hinges and near the pin hole. These results are for the PLA finger model subjected to 15 N loading. (b) Numerical results of the obtained von Mises stresses vs. applied load. (c) The endurance limit for ABS and PLA materials. Reproduced from Cabibihan et al. [9]. ©2020 Cabibihan et al. Used under Creative Commons CC BY 4.0 license.

15.5.2 Results

15.5.2.1 Material failure analysis

The results obtained show that the maximum stresses in the ABS and PLA materials have been evaluated under different loads (Fig. 15.6a). The maximum von Mises stress of 12.04 MPa was recorded for the ABS material at a load of 10 N while the stress in the PLA material was higher and reached a maximum of 44.76 MPa at a load of 25 N (Fig. 15.6b). The study employed the von Mises criterion to predict yield or fracture and concluded that failure occurs as soon as the von Mises stress is greater than or equal to the material's yield stress.

Additionally, the results show the maximum load at which the artificial fingers made of ABS and PLA materials will break at loads above 13.75 N for ABS and 28.3 N for PLA. Furthermore, the analysis shows that for the typical usage of 50 cycles of movement per day, the lifespan of the PLA artificial finger is estimated to be approximately 16 years and the ABS material is known to fail in a shorter time under high loads (Fig. 15.6c and Table 15.5).

15.5.2.2 Cable failure analysis

From the analysis, the maximum stress in the cable was found to be 100 MPa when the cable was loaded with a tension of 28 N (Table 15.6). This stress level is much smaller than the cable's nominal capacity of 1,570 MPa at a load of 440 N. Hence, it can be stated that the non-elastic cable will not break in service,

TABLE 15.5 Life cycle of the ABS and PLA prosthetic fingers with respect to the load. Reproduced from Cabibihan et al. [9]. ©2020 Cabibihan et al. Used under Creative Commons CC BY 4.0 license.

Material	Applied Load (N)				
	5	10	15	20	25
ABS	1×10^6 (no failure)	7×10^4 (4 yrs)	Fail –	Fail –	Fail –
PLA	5×10^6 (no failure)	3×10^5 (16 yrs)	7×10^4 (4 yrs)	2×10^4 (1 yr)	7.5×10^3 (0.5 yr)

TABLE 15.6 Axial stresses from the applied load on the cable. Reproduced from Cabibihan et al. [9]. ©2020 Cabibihan et al. Used under Creative Commons CC BY 4.0 license.

	Applied Load (N)					
	5	10	15	20	25	28
Axial Stress (MPa)	17.8	35.7	53.6	71.4	89.3	100.0

assuming there are no other issues with the cable, such as friction between the cable and the prosthetic material.

The findings further reveal that the cable's design and material characteristics are sufficiently strong to support the anticipated loads when the prosthetic hand is in use, making it a reliable choice.

15.6 Summary

Understanding the basic mechanism of the actuation of the human hand enables us to explore the mechanical correlates to achieve the basic tasks that we intend the human hand to accomplish. We learned that the human hand operates with extrinsic and intrinsic modalities of actuation. With extrinsic actuation, the forces produced by the hand originate from muscles in the forearm. With intrinsic actuators, the fine movements are made possible with muscles located within the hand.

For basic grasping, underactuated mechanisms have been found to be sufficient to perform basic grasping operations. The principle of underactuated mechanisms relies on the number of actuators being less than the number of DOFs of the finger movements. To compensate for the missing DOFs, passive mechanical elements like springs and linkages are used. Analogous to the tendons in the musculoskeletal system, cables are used to transmit the linear motion from an external actuator to the artificial fingers.

Analyses on an underactuated artificial hand gave practical insights into the limitations of such designs. The artificial hand was analyzed to assess the range of motion of the fingers. It was shown that the finger movements range from 0°

to 90° angles. The grasping postures include palmar pinch, inferior pincer grasp, and lateral grasp. When evaluated against the full grasp set of human hands, this sample of artificial hand can only accomplish three out of the 33 grasps. In the analysis of the material of the 3D printed hand (i.e., ABS and PLA), it was found that these samples analyzed have sufficiently long life cycles that can last for 4 years or more. The cable that was tested showed that the cable will not break in service. From the tests replicating human grasping forces, the maximum stress was found to be 100 MPa when the cable was loaded with a tensile load of 28 N. The cable's capacity is 1,570 MPa at a load of 440 N.

Knowing the anatomy and the actuation mechanisms of human hands provide us with some useful guidance in the design of the structure and the actuation mechanisms of artificial hands. By extracting some of the functional aspects of the human hand's actuation system, it would be possible to replicate some of the basic grasping tasks for an artificial hand. Underactuated mechanisms offer an optimized solution for simpler controls, reduction of weight, and minimizing the overall complexity of an artificial hand.

References

[1] J. Landsmeer, Power grip and precision handling, Annals of the Rheumatic Diseases 21 (1962 Jun) 164–170, https://doi.org/10.1136/ard.21.2.164.

[2] M.S. Han, C.K. Harnett, Journey from human hands to robot hands: biological inspiration of anthropomorphic robotic manipulators, Bioinspiration & Biomimetics 19 (2024) 021001, https://doi.org/10.1088/1748-3190/ad262c.

[3] R.S. Johansson, J.R. Flanagan, Coding and use of tactile signals from the fingertips in object manipulation tasks, Nature Reviews. Neuroscience 10 (2009 Apr) 345–359, https://doi.org/10.1038/nrn2621.

[4] F. Cini, V. Ortenzi, P. Corke, M. Controzzi, On the choice of grasp type and location when handing over an object, Science Robotics 4 (2019 Feb), https://doi.org/10.1126/scirobotics.aau9757.

[5] S. Hirose, Y. Umetani, The development of soft gripper for the versatile robot hand, Mechanism and Machine Theory 13 (1978 Jan) 351–359, https://doi.org/10.1016/0094-114x(78)90059-9.

[6] R. Deimel, O. Brock, A novel type of compliant and underactuated robotic hand for dexterous grasping, The International Journal of Robotics Research 35 (2016) 161–185, https://doi.org/10.1177/0278364915592961.

[7] C.D. Santina, C. Piazza, G. Grioli, M.G. Catalano, A. Bicchi, Toward dexterous manipulation with augmented adaptive synergies: the Pisa/IIT SoftHand 2, IEEE Transactions on Robotics 34 (2018 Oct) 1141–1156, https://doi.org/10.1109/tro.2018.2830407.

[8] F. Alkhatib, J.J. Cabibihan, E. Mahdi, Data for benchmarking low-cost, 3D printed prosthetic hands, Data in Brief 25 (2019) 104163, https://doi.org/10.1016/j.dib.2019.104163.

[9] J.J. Cabibihan, F. Alkhatib, M. Mudassir, L.A. Lambert, O.S. Al-Kwifi, K. Diab, E. Mahdi, Suitability of the openly accessible 3D printed prosthetic hands for war-wounded children, Frontiers in Robotics and AI 2021 7 (2020), https://doi.org/10.3389/frobt.2020.594196.

[10] R. Tubiana, The Hand, vol. 1, W.B. Saunders, London, England, 1979.

[11] T.A.R. Schreuders, J.W. Brandsma, H.J. Stam, Functional Anatomy and Biomechanics of the Hand. Hand Function, Springer International Publishing, 2019, pp. 3–21, https://doi.org/10.1007/978-3-030-17000-4_1.

[12] J.R. Doyle, Anatomy of the finger flexor tendon sheath and pulley system, The Journal of Hand Surgery 13 (1988) 473–484, https://doi.org/10.1016/s0363-5023(88)80082-0.

[13] S.W. Lee, D.G. Kamper, Transmission of Musculotendon Forces to the Index Finger. The Human Hand as an Inspiration for Robot Hand Development, Springer International Publishing, 2014, pp. 77–97, https://doi.org/10.1007/978-3-319-03017-3_4.
[14] L. Birglen, T. Lalibert, C.M. Gosselin, Underactuated Robotic Hands, Springer Publishing Company, Incorporated, 2008.
[15] M. Carrozza, C. Suppo, F. Sebastiani, B. Massa, F. Vecchi, R. Lazzarini, M. Cutkosky, P. Dario, The SPRING hand: development of a self-adaptive prosthesis for restoring natural grasping, Autonomous Robots 16 (2004) 125–141, https://doi.org/10.1023/b:auro.0000016863.48502.98.
[16] R. Ozawa, H. Kobayashi, K. Hashirii, Analysis classification, and design of tendon-driven mechanisms, IEEE Transactions on Robotics 30 (2014) 396–410, https://doi.org/10.1109/tro.2013.2287976.
[17] F. Alkhatib, E. Mahdi, J.J. Cabibihan, Design and analysis of flexible joints for a robust 3D printed prosthetic hand, in: 2019 IEEE 16th International Conference on Rehabilitation Robotics (ICORR), 2019, pp. 784–789, https://doi.org/10.1109/icorr.2019.8779372.
[18] T. Feix, J. Romero, H. Schmiedmayer, A.M. Dollar, D. Kragic, The GRASP taxonomy of human grasp types, IEEE Transactions on Human-Machine Systems 46 (2016) 66–77, https://doi.org/10.1109/THMS.2015.2470657.
[19] P. Corke, Robotics Vision and Control: Fundamental Algorithms in MATLAB, vol. 118, Springer, 2017.
[20] H. Wang, F.J. Abu-Dakka, T. Le Nguyen, V. Kyrki, H. Xu, A novel soft robotic hand design with human-inspired soft palm: achieving a great diversity of grasps, IEEE Robotics & Automation Magazine 28 (2021 Jun) 37–49, https://doi.org/10.1109/mra.2021.3065870.
[21] X. Gao, K. Yao, K. Junge, J. Hughes, A. Billard, A detachable crawling robotic hand, Nature Communications 17 (2026 Jan), https://doi.org/10.1038/s41467-025-67675-8.
[22] J.J. Cabibihan, M. Mudassir, Parametrically-Designed Artificial Hand with Multifunctional Grasps. Social Robotics, Springer Nature, Singapore, 2025, pp. 81–92, https://doi.org/10.1007/978-981-96-1151-5_9.
[23] United States Department of Justice, ADA Standards for Accessible Design, USA, 2010.

Chapter 16

Artificial thermo-soft hands

16.1 Introduction

Would it be possible to trick a person into believing that the hand that touches him/her is a lifelike artificial hand? If that were possible, how do we design such an artificial hand? How do we design the experiment to test such a hand with human participants?

In addition to its functional role, the human hands also have to address their social role. Social touching is that type of touch when people touch each other, when they shake hands, when they give high fives, or when one caresses the arms of a spouse [1]. A natural human-robot interaction requires that the robot demonstrate believable behavior; the robot has to follow social norms and conventions [2]. An artificial hand of a humanoid robot or a prosthetic hand will be expected to perform social touching behaviors.

This chapter presents a design and experimental methodology with the aim of making a person perceive contact from an artificial hand as if it were from human hands. In the next section, artificial skin samples were tested with touch trials with participants. An array of 16 fingertips was arranged according to increasing softness and increasing warmth.

Next, the soft-warm fingertip that was selected by the highest number of participants was implemented in an artificial hand. Two sets of hands were constructed using the imaging-3D printing-silicone casting-validation workflow described in Chap. 13. One hand included the heater, the other did not have it. The following section describes the experiment with participants who had their eyes covered and were touched by a human hand, a soft and cold hand, and a soft and warm hand.

Sec. 16.2 describes the psychophysical experiments on the fingertip. Sec. 16.3 presents the design, construction, and testing of passive artificial hands. Sec. 16.4 describes the blinded touch test where the touch from human and artificial hands were tested. Finally, Sec. 16.5 summarizes the outcomes of this chapter.

16.2 Touch trials on artificial skin samples

16.2.1 Methodology

Participants were invited to touch numerous skin samples and choose the one that felt similar to human skin tissue in order to evaluate the desired thermal and

Humanizing Artificial Touch. https://doi.org/10.1016/B978-0-32-395519-5.00028-3

mechanical properties of an artificial skin. The trials involved 95 males and 70 females, all between the ages of 17 and 28, making up the entire group of 165 healthy individuals who were chosen. Participation was voluntary.

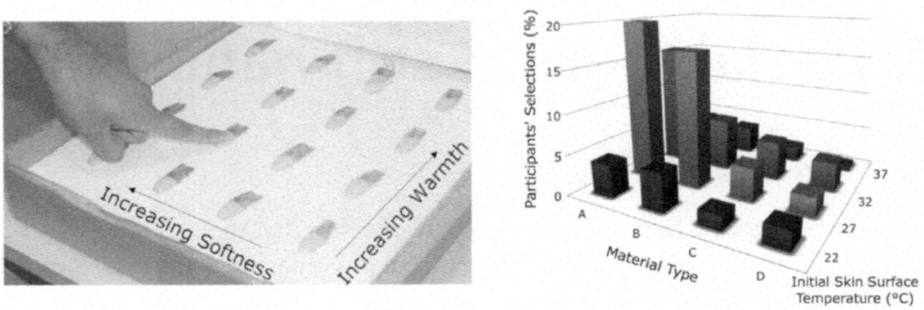

(a) (b)

FIGURE 16.1 Experimental investigations and results on tactile perception using artificial skin samples. (a) The samples were arranged in a 4 × 4 array, with the columns organized in ascending order of softness based on the materials used, and the rows arranged in ascending order of warmth. (b) The clustering of results on artificial skins with varying material types and initial surface temperatures. The preferred materials were material A with an initial surface temperature of 27°C, material A with an initial surface temperature of 32°C, and material B with an initial surface temperature of 27°C. Adapted from Cabibihan et al. [3]. ©2015 Cabibihan et al. This is an open access article distributed under the terms of the Creative Commons Attribution License (CC BY 4.0).

The objective of this study was to determine the optimal temperature for a lifelike artificial hand by evaluating the temperature preferences of the participants. It can be noted that alterations in the room temperature can impact the skin temperature of the hand of the participant, as suggested in Cabibihan et al. [4].

TABLE 16.1 Specifications of the Materials Selected. Reproduced from Cabibihan et al. [3]. ©2015 Cabibihan et al. Licensed under CC BY 4.0.

Designation	Material and Code	Manufacturer	Shore OO
A	Ecoflex, OO-30	Smooth-On	30
B	Prochima, GLS 40	Prochima	55
C	Dragon Skin, 20	Smooth-On	70
D	PDMS, Sylgard 184	Dow Corning	86

Four distinct materials with varying softness were chosen. Table 16.1 presents the specifications of the chosen materials. A lower Shore durometer value is indicative of a material that possesses a relatively lower degree of hardness. The skin samples were systematically organized to exhibit a gradual transition of soft and warm characteristics, and were positioned in a 4 × 4 matrix as depicted in Fig. 16.1a. The experimental setup was configured such that there was a consistent temperature gradient of 5°C across each row, specifically at the surface of the sample. The temperatures used were 22°C, 27°C, 32°C, and 37°C. The details for the embedded heater design can be found in an earlier work [4].

The selection of materials was determined from their previous usage in prosthetic or robotic skins [1][5], their incorporation in materials for tactile sensors [6][7], and their application in soft robotics [8][9]. The skin samples were produced utilizing conventional molding methodologies. The samples were colored with identical colors in order to eliminate any potential visual bias. The specimens were tinted with a skin-toned pigment (Silc Pig, Smooth-On, USA), which was mixed into a silicone paint base (Psycho Paint, Smooth-On, USA) to ensure proper adhesion.

16.2.2 Skin heat transfer analysis

A change in the temperature occurs at the surface when the human fingertip touches the skin material sample. The temperature change can be determined when the thermal characteristics of the four materials are known. To measure the thermal conductivity of each material, a thermal conductivity analyzer (TCi, C-Therm Technologies, Canada) was used. The density of each material was obtained from the data sheets. The specific heat can then be calculated. The thermal conductivity, density, and specific heat for each material are shown in Table 16.2.

TABLE 16.2 Material Properties of Soft Materials. Reproduced from Cabibihan et al. [3]. ©2015 Cabibihan et al. Licensed under CC BY 4.0.

Material	Thermal Conductivity, k (W/m·K)	Density, ρ (kg/m^3)	Specific Heat, c (J/kg·K)
A	0.15	1,065	1,558
B	0.23	1,123	1,244
C	0.16	1,081	1,435
D	0.12	1,040	1,610

Upon entering the experiment room, the participants' initial skin temperatures were recorded. Subsequent measurements were taken at intervals of 10, 20, and 30 minutes after their arrival. A total of 30 individuals (comprising 15 males and 15 females, all aged between 18 and 25) were randomly chosen from those involved in the skin sample studies. The participants were seated comfortably in a relaxed posture. The ambient room temperature was kept steady at 21°C. Skin temperatures were measured at the papillary whorl of the index fingertip on the participant's right hand, using an infrared thermometer (model 62 Mini, Fluke, USA; with an accuracy of ±1°C). All measurements were performed with the thermometer held at a fixed distance of 22 cm. The average temperature at the fingertip upon room entry was 28.9°C, with a standard deviation of 2.2°C.

The majority's choice was consistent with the result of the heat transfer process that occurs when one touches a surface. Upon contact, the heat from the human fingertip was transferred to the surface of the artificial fingertip by conduction. Initial temperatures of the human and artificial fingertips and their thermal properties (i.e., thermal conductivity, density, specific heat) determine the surface temperature at the contact interface. The surface temperature of each material was calculated according to [10][11]:

$$T_s = T_{artif} + \frac{(T_{human} - T_{artif})\sqrt{(k\rho c)_{human}}}{\sqrt{(k\rho c)_{human}} + \sqrt{(k\rho c)_{artif}}} \tag{16.1}$$

where T_s denotes the surface temperature when the human finger touches the artificial finger, T_{human} is the human fingertip's initial temperature, T_{artif} is the artificial fingertip's initial temperature, k is the thermal conductivity, c is the specific heat, and ρ is the density. The average initial temperature of the participants' fingertips was recorded as 29.0°C.

With $\sqrt{k\rho c}$ being the thermal contact coefficient, the value for the human fingertip was estimated to be 1,181 J/$(m^2 s^{1/2} K)$ [12]. The properties of the artificial skin materials, including their thermal conductivity, density, and specific heat, were determined through experimental measurements, as detailed in Table 16.2.

For the model to be valid, the Fourier number, F_0, must be less than 5×10^{-2} to ensure that the semi-infinite model assumption is satisfied [13]. The Fourier number is given as:

$$F_0 = \frac{\alpha t}{L_c^2} \tag{16.2}$$

where contact time, t, was set to 5 s, and the characteristic length, L_c, is the material volume over the contact area and was set to 4×10^{-3} m. Thermal diffusivity, α, was calculated as $\alpha = k(\rho c)^{-1}$. The Fourier number (Eq. 16.2) for each material was found to be approximately 5.1×10^{-2}. While this value marginally exceeds the strict criterion of 5.0×10^{-2}, it is sufficiently close to validate the use of the semi-infinite solid approximation for the brief contact times in this study.

16.2.3 Results

The samples depicted in Fig. 16.1b were chosen by the participants. The probability of selecting a sample is 6.25%, which can be expressed as 1 out of 16. In terms of softness, it was found that 46% of the participants exhibited a preference for material A, whereas 32% of the participants favored material B. In regard to temperature, it was found that 45% of the participants selected a skin surface temperature of 27°C, while 31% opted for a temperature of 32°C. It was observed that the inclinations of the participants were concentrated on three categories, namely material A with a temperature of 27°C ($n = 36$), material A with a temperature of 32°C (n = 26), and material B with a temperature of 27°C (n = 28). The temperature with the highest preference among the participants was 28.4°C based on proportional analysis. To summarize, the soft material that was most favored is material A (Ecoflex OO-30). The proportion of highly selected temperatures suggests that the optimal temperature can be established at 28.4°C.

With reference to Fig. 16.1b, the surface temperatures that correspond to the preferred softness and temperature combinations were calculated in Table 16.3. These values are notably close to the average initial fingertip temperature of the participants, which was 29.0°C and is close to the optimal temperature.

TABLE 16.3 Skin Surface Temperature for the Most Preferred Softness-Temperature Combinations.

Material	Temperature (°C)	Skin Surface Temperature (°C)
A	27	28.4
B	27	28.3
A	32	29.9

16.3 Development of a passive, lifelike artificial hand

16.3.1 Methodology

The study's participant, a 35-year-old man, was recruited for the development of the passive, lifelike artificial hand. His dominant right hand's measurements were used to make a silicone rubber hand replica. The participant's hand measures 19 cm in length from the crease at the wrist to the tip of the middle finger and 9 cm in width when measured across the ends of the metacarpal bones. The participant was chosen because the anthropometric data for his hand was near the 50th percentile in [14]. The Institutional Review Board provided the approval for the experiments.

Two artificial hand prototypes were created to conduct a comparative analysis of the tactile softness between the human hand and silicone rubber replicas. The term "softness" is operationally defined as the perceptual manifestation of skin compliance, which refers to the extent of deformation induced by an exerted force [15]. Both artificial hands were fabricated using the data obtained from the right hand of the human participant. The initial design employed a skeletal framework, whereas the subsequent one did not incorporate such a structure. The artificial hand with bones (WB) and the artificial hand without bones (NB) shall be referred to as such, with the former featuring a skeleton and the latter lacking one. Fig. 16.2 shows the human hand and the artificial hands that were produced. The synthetic skin material utilized in the study was Ecoflex OO-30, as selected by the experiment participants. The detailed description of the fabrication process can be found in [16].

Fig. 16.3a depicts the experimental arrangement utilized for the indentation examinations. The robotic arm utilized in the study is the IRB-140 model (ABB

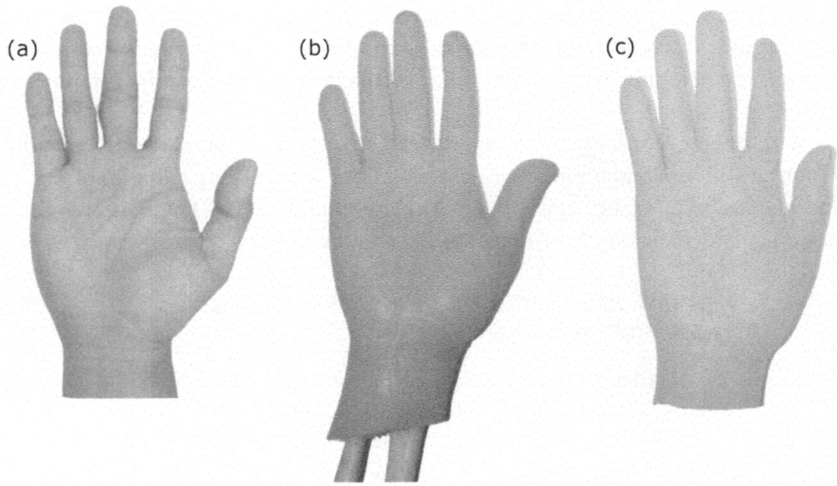

FIGURE 16.2 Human and artificial hands. (a) The experiment participant's hand. (b) Replica of the human hand with silicone material and 3D printed bone structure. (c) Silicone hand without the bone structure.

Robotics, Sweden). The experimental procedure involved placing the hand in a supine position. The study conducted indentation tests on predetermined hand regions to generate softness maps for the purpose of comparing the human and artificial hands.

16.3.2 Results

The indentation test from a probe on the robot arm helps to create the softness map for both the human and artificial hands. This map was constructed from the skin displacement data when several data points were collected at the palm-side of the hand. The data was then overlaid on the hand to show the locations that are soft or stiff.

When a constant force of 1 N was indented on the human and artificial hands, the softness map (Fig. 16.3b) shows the locations where the displacements of 0.2 to 5.75 mm were reached. The locations of the human hand where joints were located displayed the least displacement values, and the pulp locations show higher displacement values.

The softness map of the artificial hand with a skeleton structure is shown in Fig. 16.3c. The artificial hand in the no-bones (NB) condition demonstrated displacements ranging from 2.5 to 7.5 mm, whereas the artificial hand in the with-bones (WB) condition displayed displacements ranging from 0.5 to 6 mm, which were closer to the skin compliance of the human hand.

The displacements of the human hand and the artificial hands are compared in Fig. 16.3d. The data consistently indicates that the displacement of the human hand is the lowest, with the artificial hands of WB and NB following closely

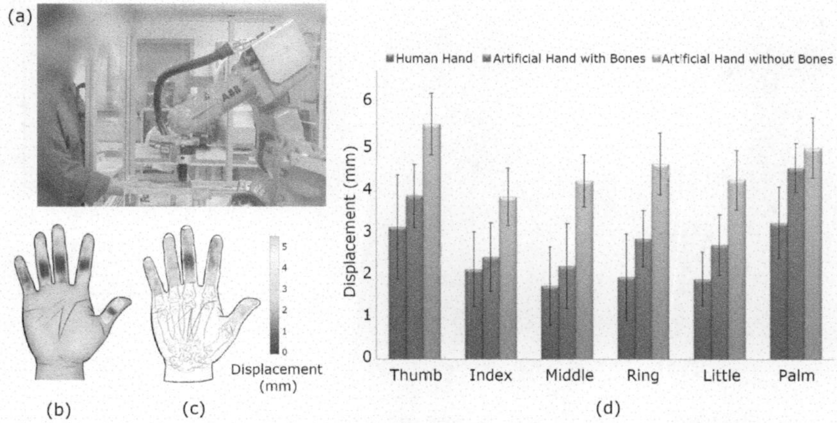

FIGURE 16.3 Indentation results on the human hand. (a) A constant force of indentation was exerted on the hand by the robotic arm. A force sensor was employed to measure the resultant displacement. (b) The skin softness maps through indentation tests conducted with a force of 1 N on the human hand and (c) on an artificial hand with bone structure (WB). (d) The skin displacements observed at a constant force of 1 N for the human hand, an artificial hand with bones (WB), and an artificial hand without bones (NB). The mean and standard deviation (SD) were calculated for each condition. The results of post-hoc analyses indicate statistically significant differences ($p < 0.001$) among all groups in relation to each of the fingers and the palm.

behind. Upon comparison of average displacements, it was observed that the artificial hand with bones exhibited a variance ranging from 14% to 47% in comparison to the human hand. In contrast, the artificial hand without bones displayed a variance ranging from 54% to 143% in comparison to the human hand for each finger and the palm. Collectively, the boneless prosthetic hand exhibited the lowest degree of rigidity compared to the other two hand models, owing to the lack of a skeletal framework. The recorded displacements for the artificial hand with bones exhibited a greater proximity to those of the human hand.

16.4 Trials on human and artificial hands

16.4.1 Methodology

The aim of these experiments is to compare the participants' responses after the touch of a hand at the hairy region of the forearm. Three distinct types of hands were utilized for the experiment, namely: a human hand, a soft and cold artificial hand, and a soft and warm artificial hand with a heating design. Both artificial hands included a 3D printed skeleton. The study recruited 28 healthy participants, comprising 20 males and 8 females, aged between 17 and 26 years old. Participation was voluntary. The study's participants were provided with instructions to discern whether the hand that made contact with them was from a human or an artificial touch.

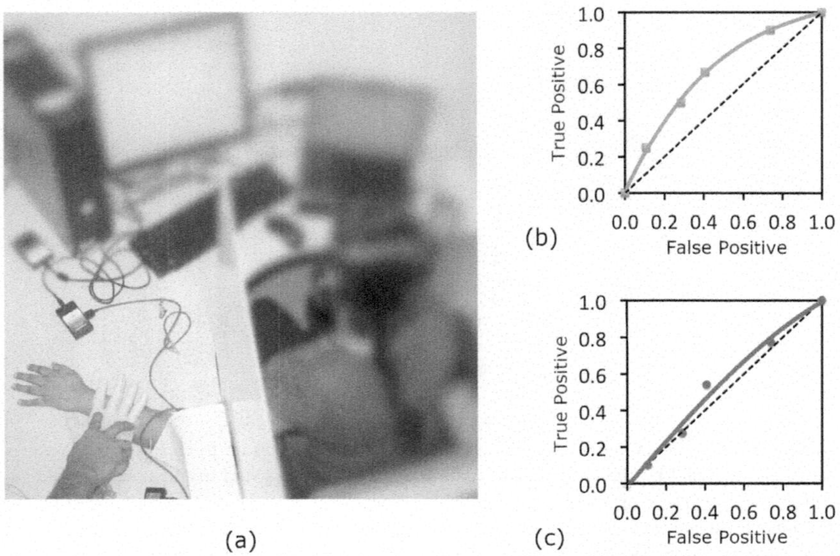

FIGURE 16.4 Experiments involving perception of human and artificial hands with bones and soft and warm artificial skins. (a) The experimental configuration involved physical contact between the test hands and the study participants. (b) The Receiver Operating Characteristic (ROC) curve analysis of the soft and cold artificial hand in comparison to the human hand. (c) The ROC curve for the artificial hand with soft and warm skin when compared to that of the human hand. Adapted from Cabibihan et al. [3]. ©2015 Cabibihan et al. This is an open access article distributed under the terms of the Creative Commons Attribution License (CC BY 4.0).

The artificial hand, which was both soft and cold, lacked a heating mechanism, resulting in its skin surface temperature equilibrating to the ambient temperature of 21°C. The artificial hand, which was designed to be both soft and warm, was initially programmed with a skin surface temperature of 28.4°C. The study's participants were unaware of both the experiment's hypothesis and the specific types of hands that would be used in the tactile stimulation. Throughout the experiment, the participants were not provided with a view of the hands, as depicted in Fig. 16.4a.

The participants were requested to provide their feedback on a 5-point Likert scale, with 1 indicating complete certainty that the tactile sensation was produced by an artificial hand and 5 indicating complete certainty that the tactile sensation was produced by a human hand (Table 16.4). The participants utilized a computer mouse to input their responses, which were measured on a 5-point scale. The study employed a pseudorandom presentation of the three hand types, with each type being equally represented across 27 total trials. To mitigate the impact of practice effects, the initial 6 trials were excluded from the analysis. A set of 588 responses was acquired, comprising 28 subjects and 21 trials per subject. The dataset proved to be adequate, as it has been recommended that a minimum of 100 observations is required for receiver operating characteristics (ROC) curve analysis in order to arrive at a significant conclusion [17].

TABLE 16.4 Frequency of the Participants' Responses for the Touch Test. Reproduced from Cabibihan et al. [3].

Hand Type	Participants' Choices				
	Very Certain to be Prosthetic	Certain to be Prosthetic	Neutral	Certain to be Human	Very Certain to be Human
	(1)	(2)	(3)	(4)	(5)
Human	21	35	24	65	51
Cold	49	49	33	46	19
Warm	20	34	52	45	45

The ROC curve analysis was used to relate the proportion of correctly recognized touches by the warm or cold artificial hand (true positive) to the proportion of mistakenly recognized touch from the human hand (false positive). The differences between the two Areas Under the Curves (AUC) were compared with the method of Hanley and McNeill [18]. In this method, the p-value was determined as the probability that the observed sample AUC is found when the true population AUC is 0.5 and that the null hypothesis assumes that the area is 0.5. The probability selected as significant was $p < 0.05$.

16.4.2 Results

Figs. 16.4b and 16.4c show the ROC results of the participants. An outcome that lacks the ability to differentiate between accurately and erroneously identified touches is represented by a ROC curve that coincides with a diagonal line and encompasses an AUC value of 0.5. Stated differently, an AUC value in proximity to 0.5 indicates that the sensation elicited by an artificial hand is indistinguishable from that of a human hand.

The results depicted in Fig. 16.4b indicate that the ability to distinguish between the touch of a soft and cold artificial hand and that of a human hand was observed ($AUC = 0.697$, $p < 0.0001$). By contrast, it can be inferred from Fig. 16.4c that the participants encountered challenges in accurately identifying the artificial hand's softness and warmth in comparison to the human hand ($AUC = 0.562$, $p = 0.0611$). The statistical analysis revealed a p-value of less than 0.05 for the cold hand, which suggests that the area under the curve (AUC) is significantly distinct from a value of 0.5. The statistical analysis of the warm artificial hand revealed a p-value exceeding 0.05, indicating that the tactile sensation elicited by this device is indistinguishable from that of a human hand.

16.5 Summary

Understanding the social requirements of artificial hands for robotics and prosthetic hands for social touching (e.g., handshake, patting, caress) adds to the layer of design considerations that we need to make in addition to the functional tasks (e.g., grasping and manipulation). For a simple touching task with a slow pressing motion, we were able to identify that the material softness and warmth of an artificial hand are important variables to consider.

An artificial lifelike hand was achieved with a three-part procedure. First, the participants were instructed to choose the artificial skin samples that had some resemblance to human skin from an array of 16 samples arranged in a 4 $\times$ 4 matrix. The majority of participants opted for a silicone material with a Shore durometer value of 30 on the OO scale. The selected temperatures were concentrated at 27°C and 32°C, within an ambient room temperature of 21°C. Upon calculation, the temperature selected was determined to be 28.4°C.

Secondly, it was demonstrated that it is possible to imitate the softness of human hand skin tissues by reproducing the surface topology of the hand and the geometries of its skeleton. This was done by utilizing CT scanning, computer-aided design, 3D printing, and silicone molding techniques. Lastly, the preceding outcomes were combined to develop the artificial hand design for the purpose of conducting perception trials. Participants received touch from a human hand and two artificial hands. The study involved covering the eyesight of the participants and asking them to differentiate between tactile sensations originating from a human hand and those from the artificial hands (i.e., soft and cold hand and soft warm hands). According to the ROC curve analysis, it has been shown that the embedded heaters on soft synthetic skin and a skeleton structure in an artificial hand can potentially create an illusion of human touch (AUC $= 0.562$; $p = 0.0611$).

The current chapter confirms that it is possible to trick the experiment participants into believing that the touch from an artificial hand is human touch. The three-step procedure described in this chapter allowed us to test that hypothesis. The psychophysics experiments on the 4 $\times$ 4 array of soft-warm finger preferences and the blindfold test with human and artificial hands further provide a framework to evaluate whether artificial hands can create a believable lifelike touch similar to human touch.

References

[1] J.J. Cabibihan, S. Pattofatto, M. Jomaa, A. Benallal, M.C. Carrozza, Towards humanlike social touch for sociable robotics and prosthetics: comparisons on the compliance, conformance and hysteresis of synthetic and human fingertip skins, International Journal of Social Robotics 1 (2009) 29–40, https://doi.org/10.1007/s12369-008-0008-9.

[2] T. Fong, I. Nourbakhsh, K. Dautenhahn, A survey of socially interactive robots, Robotics and Autonomous Systems 42 (2003) 143–166, https://doi.org/10.1016/s0921-8890(02)00372-x.

[3] J.J. Cabibihan, D. Joshi, Y.M. Srinivasa, M.A. Chan, A. Muruganantham, Illusory sense of human touch from a warm and soft artificial hand, IEEE Transactions on Neural Systems and

Rehabilitation Engineering 23 (2015) 517–527, https://doi.org/10.1109/tnsre.2014.2360533.
[4] J.J. Cabibihan, R. Jegadeesan, S. Salehi, S.S. Ge, Synthetic Skins with Humanlike Warmth, Lecture Notes in Computer Science, Springer, Berlin, Heidelberg, 2010, pp. 362–371, https://doi.org/10.1007/978-3-642-17248-9_38.
[5] J.J. Cabibihan, F. Alkhatib, M. Mudassir, L.A. Lambert, O.S. Al-Kwifi, K. Diab, E. Mahdi, Suitability of the openly accessible 3D printed prosthetic hands for war-wounded children, Frontiers in Robotics and AI 7 (2021), https://doi.org/10.3389/frobt.2020.594196.
[6] D.H. Kim, J. Viventi, J.J. Amsden, J. Xiao, L. Vigeland, Y.S. Kim, J.A. Blanco, B. Panilaitis, E.S. Frechette, D. Contreras, D.L. Kaplan, F.G. Omenetto, Y. Huang, K.C. Hwang, M.R. Zakin, B. Litt, J.A. Rogers, Dissolvable films of silk fibroin for ultrathin conformal bio-integrated electronics, Nature Materials 9 (2010 Apr) 511–517, https://doi.org/10.1038/nmat2745.
[7] S.C.B. Mannsfeld, B.C.K. Tee, R.M. Stoltenberg, C.V.H.H. Chen, S. Barman, B.V.O. Muir, A.N. Sokolov, C. Reese, Z. Bao, Highly sensitive flexible pressure sensors with microstructured rubber dielectric layers, Nature Materials 9 (2010 Sep) 859–864, https://doi.org/10.1038/nmat2834.
[8] R.F. Shepherd, F. Ilievski, W. Choi, S.A. Morin, A.A. Stokes, A.D. Mazzeo, X. Chen, M. Wang, G.M. Whitesides, Multigait soft robot, Proceedings of the National Academy of Sciences 108 (2011 Nov) 20400–20403, https://doi.org/10.1073/pnas.1116564108.
[9] F. Tramacere, A. Kovalev, T. Kleinteich, S.N. Gorb, B. Mazzolai, Structure and mechanical properties of octopus vulgaris, Journal of the Royal Society Interface 11 (2014 Feb) 20130816, https://doi.org/10.1098/rsif.2013.0816.
[10] J.H. Lienhard IV, V.J.H. Lienhard, A Heat Transfer Textbook, Phlogiston Press, Cambridge, MA, 2008 Apr.
[11] L. Jones, H.N. Ho, Warm or cool, large or small? The challenge of thermal displays, IEEE Transactions on Haptics 1 (2008 Jan) 53–70, https://doi.org/10.1109/toh.2008.2.
[12] H.N. Ho, L.A. Jones, Modeling the thermal responses of the skin surface during hand-object interactions, Journal of Biomechanical Engineering 130 (2008), https://doi.org/10.1115/1.2899574.
[13] H.N. Ho, L.A. Jones, Contribution of thermal cues to material discrimination and localization, Perception & Psychophysics 68 (1) (2006) 118–128, https://doi.org/10.3758/bf03193662.
[14] Human Factors Engineering Technical Advisory Group, Human Engineering Design Data Digest, Department of Defense, 2000.
[15] R.M. Friedman, K.D. Hester, B.G. Green, R.H. LaMotte, Magnitude estimation of softness, Experimental Brain Research 191 (2008 Aug) 133–142, https://doi.org/10.1007/s00221-008-1507-5.
[16] J.J. Cabibihan, M.K. Abubasha, N. Thakor, A method for 3-D printing patient-specific prosthetic arms with high accuracy shape and size, IEEE Access 6 (2018) 25029–25039, https://doi.org/10.1109/access.2018.2825224.
[17] N.A. Macmillan, C.D. Creelma, Detection Theory: A User's Guide, Taylor and Francis, 2005.
[18] J.A. Hanley, B.J. McNeil, The meaning and use of the area under a receiver operating characteristic (ROC) curve, Radiology 143 (1982 Apr) 29–36, https://doi.org/10.1148/radiology.143.1.7063747.

Part V

Conclusions

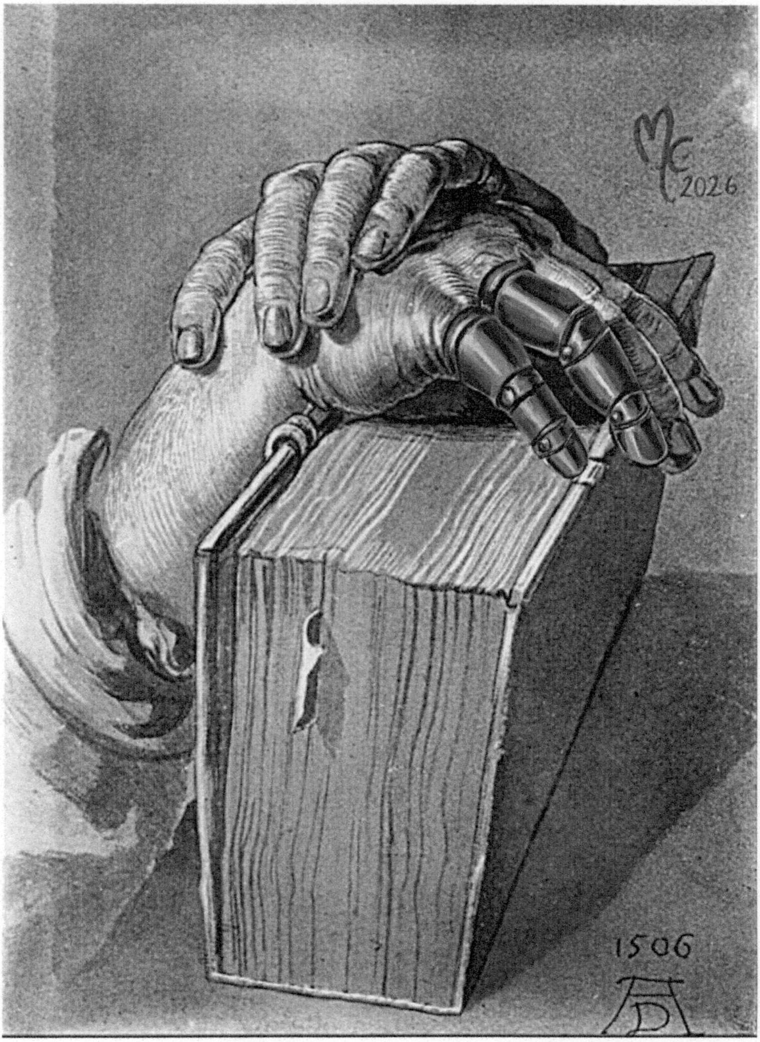

Chapter 17

Conclusions

17.1 Concluding remarks

From our first cry to our last breath, the human touch has been the universal language that binds families, builds societies, and defines our shared humanity. Touch communicates love, empathy, and trust; intimacy and attraction; hierarchy and power; care and healing. But what if we remove that human touch?

For those with amputations from war or accident, the loss is profound. This absence can lead to depression, anxiety, and a recurring sense of loneliness. The absence of touch creates immense functional and psychosocial consequences. Conversely, in human-to-robot touch interactions, the touch from a cold, stiff humanoid robot hand could result in a sense of inauthenticity or dissonance. Genuine human-to-human interaction releases oxytocin, or the bonding hormone, and its absence makes the experience cold, robotic, and artificial.

This book proposes a bio-inspired and integrated systems approach to replicate a lifelike touch experience. The core thesis is that by understanding the nuanced interplay between the skin, tactile sensing, and the human hand, we can design artificial hands that are not only functionally proficient but also capable of believable social interactions.

The book begins by elaborating the two-part framework of human touch [1]. First, when humans see an object, the internal models and action programs are activated to anticipate the object's properties, like weight and texture, and pre-plan the necessary grip forces (i.e., the Anticipatory Predictive Control). Second, once the fingers make contact, the fingertip skin and mechanoreceptors provide immediate feedback. This updates the internal models and triggers compensatory actions for any errors between the predicted and actual sensory input (i.e., Discrete Event Sensory Driven Control). For robotic systems, this model translates into an integrated feedforward-feedback control architecture. A vision module provides initial object information to the robot's internal model, while sensors in the robot hand provide real-time contact information for rapid, corrective feedback.

The book adopts a bottom-up framework that strongly emphasizes that the design should start from the contact interface with an object, which is the soft skin, and progress upwards to the tactile sensors and the hand structure. This

Humanizing Artificial Touch. https://doi.org/10.1016/B978-0-32-395519-5.00030-1

three-part architecture highlights the mutual contributions of each component in an attempt to achieve human-like artificial touch.

17.2 Key contributions

The book provides a conceptual framework and a practical road map for creating artificial hands that feel and interact with some resemblance to human hands. It combines a component-level focus and the systems-level integration of the skin, sensing, and structures toward achieving the grand challenge of humanizing artificial touch.

17.2.1 Part II: Skin

Part II established the foundational importance of the contact interface. As demonstrated in Chapters 2 through 6, the properties of the skin, such as conformance, compliance, hysteresis, relaxation, and creep are not passive but active features that have a significant effect on the response of the embedded sensors. Modeling and replicating such features elevate the skin from simple external covers to its new role in providing mechanical intelligence to the embedded sensors.

- By characterizing the hyperelastic and viscoelastic properties of materials like silicone and polyurethane (Chap. 3), the book demonstrates how Finite Element Analysis can be used to predict the mechanical behavior of the skin before physically embedding sensors. This insight moves from the costly trial-and-error approach to optimizing the material selection and geometry at the outset (Chap. 4).
- Investigating the skin's time-dependent behavior (i.e., hysteresis, relaxation, creep) gave us an understanding of the drift and delayed recovery of tactile sensing signals (Chap. 5). This insight makes it clear that embedded tactile sensors must either use materials with minimal time-dependent effects or use a control strategy that compensates for the errors.
- The skin thickness is a critical design parameter; the skin functions as a low-pass filter (Chap. 6). This insight forces the designer to consider the trade-off between sensor acuity and durability. Thinner skins preserve the high-frequency signals while thicker skins blur out the details.

17.2.2 Part III: Sensing

Part III investigated some of the basic requirements for tactile sensing for grasping. As shown in Chapters 7 through 11, the bio-inspired approach to sense contact, slippage, and shape has transformed the device-level on-off contact sensors, triaxial force sensors, optical sensors, and flexure sensors into sensing systems that can replicate some of the most basic functionalities of the human tactile sensing system and be used to achieve a more adaptive sensory feedback loop.

- The book shifts the paradigm of artificial sensing from the traditional data transduction to bio-inspired information processing. The insights about the specialized roles of different human mechanoreceptors (i.e., SA I, FA I, SA II and FA II) allow us to transform off-the-shelf sensors to tactile sensor systems that can effectively replicate the encoding of complex tactile signals for contact, slippage, and local shape (Chap. 7–10).
- The insight into incipient versus full-slip detection aids in the development of a slippage sensing system (Chap. 9). The book describes a control strategy that detects the localized slips, suppresses them and implements a reflex-like response.
- Similar to proprioceptive sensors, the usage of flexure sensors into the finger joints presents a practical way to infer the overall shape of a grasped object (Chap. 11). This insight was helpful in discriminating between various primitive shapes through flexure sensors without the need for a complex array of tactile sensors.

17.2.3 Part IV: Structures

Finally, Part IV integrated those principles into the physical embodiment of the artificial hand. Chapters 12 through 16 showed that the human hand's dexterity is influenced by the musculoskeletal system. The design elements of artificial hands, consisting of the structural, transmission, and actuation systems can take inspiration from the human bones, ligaments, and tendons. The CT scan to silicone casting workflow is an enabling technique for lifelike structures.

- A major design contribution is the streamlined, digital workflow for creating personalized artificial fingers and hands (Chap. 13). By integrating CT scanning, CAD mirroring techniques, 3D printing of molds, and silicone casting, this workflow solves the challenge of producing anatomically accurate artificial hands.
- The analysis of joints and underactuated mechanisms provides a clear insight and guidance for robot grasping (Chap. 14–15). This insight simplifies the control problem significantly, enabling a robust and adaptive grasp without requiring complex sensor feedback for every joint.
- An optimal softness and warmth and integrating these into an anatomically correct hand have been identified (Chap. 16). The finding that a soft, warm artificial hand is statistically indistinguishable from a human hand in a blind touch test suggests that an artificial hand can be perceived as human.

17.3 Limitations

The focused scope of this book gives rise to several limitations, which represent important opportunities for future research.

- The book experimented only on two types of soft materials: silicone and polyurethane. There are other materials, like hydrogels, composites, self-healing materials, that were not explored.
- The novel smart e-skins offer wide-area sensing integrated in a thin, flexible form factor. The book addressed the skin and sensors separately. Future systems will involve an integrated system of thin-film materials, sensing, electronics, and computational capability.
- The active thermal feedback systems of the artificial skin were not explored. The skin temperature used was restricted to ambient room temperature. The aesthetics on the skin coloration, blemishes, and hair on the skin surface were also not explored. The skin temperature and coloration are different under lab conditions and under different weather conditions.
- The book draws inspiration from the human neurophysiology but does not attempt to model its full complexity. The control strategies and machine learning approaches presented were simplified proof-of-concept attempts. The higher-level cognitive and emotional aspects of touch, relating to the anticipatory feedforward control, memory, context, and affective states in the brain, among others, are outside of the scope of this introductory text.
- The human hand is capable of 33 grasp poses. The case study on underactuated artificial hand presented here was only able to accomplish 3 grasp types. Other robot or prosthetic hands are now able to achieve most of the 33 grasps. The long-term durability of all these artificial hands needs to be evaluated functionally and socially in the real-world.

17.4 Outlook

17.4.1 Turing's test

A machine's ability to emulate human capabilities has been the subject of a test proposed by Alan Turing in the 1950s paper titled Computing Machine and Intelligence [2]. Turing devised a game which asked "Can machines think?".

17.4.1.1 Turing's imitation game

Turing named the game as the Imitation Game. The game involves three participants: a man designated as A, a woman as B and an interrogator as C. The interrogator is separated from the other two participants and communicates with them through text-only interface. This eliminated auditory cues like the voices of the participants. The interrogator's goal is to determine which of the two hidden individuals is the man and which is the woman.

As part of the game, the man (A) attempts to trick the interrogator into misidentifying him as the woman (B). Player A's strategy is to deceive the interrogator to make the wrong identification. The woman (B) aims to help the interrogator to make the correct identification, by giving truthful answers.

Turing then raised the question: "What will happen when a machine replaces player A in this game?". The machine now takes player A's role and attempts to imitate a human. In this variation, the machine is trying to pass as a human woman to trick the interrogator who is trying to distinguish between a human man and a human woman. The interrogator is now communicating with the human (B) and the machine (A) and determines which is which. The machine passes the test if the interrogator (C) mistakenly identifies the machine as human as often as the interrogator misidentifies a human when the game is played between two humans. This implies that the machine's responses are so human-like that an average interrogator cannot reliably distinguish them from those of a human.

Turing provided examples of questions that an interrogator might ask and the machine's response to make its answer appear as human.

> ***Interrogator:*** *Add 34957 to 70764.*
> ***Machine (acting as a human):*** *(Pauses for 30 seconds, then answers.) 105621.*
> *The machine introduces a human-like delay, reflecting that an average human would not add that quickly. An intentional error was also made, with the correct answer being 105721.*

Turing predicted that by the end of the 20th century, it would be possible to program computers that an average interrogator would not have more than a 70% chance of answering correctly after five minutes of questioning.

17.4.1.2 AI's performance on the Turing test

Recent developments in large language models (LLMs) have shown significant advancements in thinking machines. In a 2025 study by Jones et al. [3], they investigated whether people could distinguish modern AI systems from humans. Their randomized, controlled Turing test involved human participants in a 5-minute conversation with either a human or an AI. In their study, OpenAI's GPT-4 model was judged to be human 54% of the time. Because this rate is statistically indistinguishable from chance (50%), the authors concluded that GPT-4 passed the Turing Test.

A replication study [3] used GPT-4o with a "tiktok" prompt that instructed the model to adopt a specific persona with slang and occasional spelling errors. In this test, GPT-4o achieved a pass rate of 77%, which was higher than the pass rate of human participants of 71% in the same study. These results provide strong evidence that a machine has successfully passed a modified 2-person version of the Turing Test, exceeding the 30% pass rate benchmark originally suggested by Turing [2].

17.4.1.3 Turing test and humanizing artificial touch

With the recent success of LLMs in the linguistic Turing Test in imitating human conversations, this book envisions a more challenging humanization that includes the nuances of physical interaction with touch. The book proposes a touch test as a test platform to benchmark touch interactions. This version of the touch test requires materials science, haptics, and robot control that aims to achieve an artificial touch that is indistinguishable from human touch. Beyond this, the robot's interaction partner must evaluate whether the human-like artificial touch has an emotional effect.

17.4.2 Mori's uncanny valley

The Uncanny Valley, introduced in 1970 by Masahiro Mori [4][5], described a peculiar relationship between an artifact's similarity to humans and the emotional response it elicits from an observer.

17.4.2.1 The uncanny valley hypothesis

Based on Mori's observation, a robot designed with little similarity to humans, like an industrial robot, elicits a bland response from a human observer. Increasing an artifact's physical similarities to a human, but with less humanlike functionalities makes those artifacts more appealing (Fig. 17.1). In the first peaks of the dynamic and static artifacts, we can see examples like a humanoid robot or a teddy bear.

Making an artifact look and function more like human, makes that artifact more revolting and disgusting to the observer. Mori classified prosthetic hands to

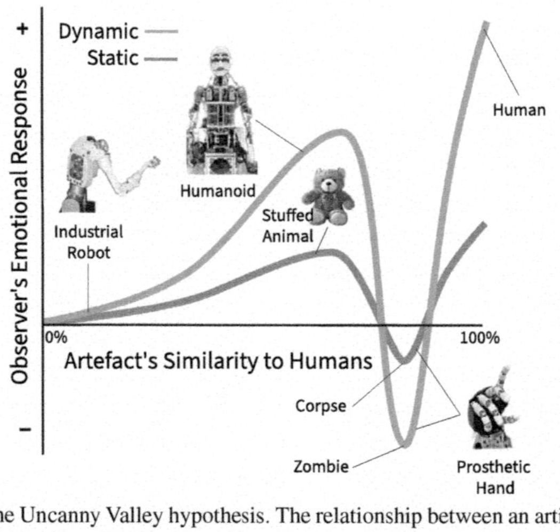

FIGURE 17.1 The Uncanny Valley hypothesis. The relationship between an artifact's similarity to humans and the human observer's emotional response to various dynamic and static artifacts [4][5]. Reproduced from Cabibihan et al. [6], with permission. ©2006 IEEE.

be at the bottom of the valley. The cosmetic prosthetic hands in the static curve and the electromyogram (EMG)-controlled prosthetic hands in the dynamic curve were designated near the non-moving corpses and the moving zombies at the valley. According to Mori, if a cosmetic prosthetic hand were to be shaken, as in a handshake gesture, the other person will be surprised by the lack of soft tissue and warmth. If an EMG-controlled hand were to be shaken, the prosthesis user will be placed in a state of shock due to the EMG.

To create a more elegant and fashionable prosthetic hand, Mori further suggested that the designs should be made of wood, to keep the natural wooden color, and ignore the fine features like the fingerprints.

17.4.2.2 Contradictions on prosthetic hands

While the Uncanny Valley may have been applicable for the faces of humanoid robots, for computer-generated imagery (CGI) in movies, animation and games, and for immersive technologies in virtual and augmented reality, we directly challenged Mori's recommendations on prosthetic hands in our 2006 paper [6]. We argued that the reasons he cited in 1970 in using wood, wood-coloring, and fingerprints were not feasible in the context of the contemporary design trends and the psychosocial needs of the amputees. For skin material, we argued that soft skins are crucial for robotic or prosthetic grasping and manipulation. For skin color, amputees do not want to attract unwanted attention. For fingerprint features, we know that fingerprints have functional benefits for tactile signal amplification, adhesion, and slippage sensing. Back then, we concluded that the Uncanny Valley hypothesis was not applicable for prosthetic hands.

Subsequent empirical studies supported the notion that instead of avoiding realism, the humanized appearance to prosthetic hands can lead to better acceptance. In a 2013 study by Poliakoff et al. [7], they provided the first empirical demonstration that prosthetic hands with intermediate human-likeness were rated as more eerie than mechanical or real hands. As the human-likeness increased, they observed a reduction in eeriness. In their 2018 follow-up study [8], they found that less-realistic prosthetic hands consistently generate a sense of eeriness on the observers on initial viewing, while more realistic prosthetic hands avoid the Uncanny Valley.

17.4.2.3 Uncanny valley and humanizing artificial touch

The premise of this book is to create artificial hands that are bio-inspired and human-like in some of the important characteristics of the skin, sensors, and structures to create emotional connection and natural human-robot interaction. This objective directly disagrees and contradicts Mori's original hypothesis for prosthetic hands where non-humanized designs are to be adopted to avoid the Uncanny Valley. Over the years, the Uncanny Valley has been found to be applicable in facial appearances and movements. This book presents detailed framework and empirical studies to navigate and overcome the valley for tactile experiences for prosthetic and robotic hands.

17.4.3 Brooks' laws of robotics

17.4.3.1 Brooks' three laws

As robots move out from the laboratories to the real world, deploying robots face its corresponding challenges and expectations. In 2024, Rodney Brooks, Professor Emeritus at MIT and cofounder of iRobot, Rethink Robotics, and Robust AI, gave an insightful perspective on the deployment of robots, emphasizing reliability, user acceptance, and practical utility [9].

1. **The Promise Given by Appearance.** The visual appearance of a robot makes a promise about what it can do and how smart it is. It needs to deliver or slightly over-deliver on that promise or it will not be accepted.
2. **Preserving People's Agency.** When robots and people coexist in the same spaces, the robots must not take away from people's agency, particularly when the robots are failing, as they inevitably will at times.
3. **Robots that Work Every Time.** Technologies for robots need 10+ years of steady improvement beyond lab demos of the target tasks to mature to low cost and to have their limitations characterized well enough that they can deliver 99.9% of the time. Every 10 more years get another 9 in reliability.

The first law warns against creating robots that appear to be more capable than they are because this leads to customer disappointment and lack of adoption. For the iRobot Roomba, its disk-shape appearance was made intentional to be a floor vacuum. For rough-terrain, the iRobot Packbot was designed to look like a mini-tank.

The second law states that robots should not obstruct or interfere with human activities, especially in critical situations. Brooks provided examples of hospital robots of not having situational awareness and blocking the nurses during emergency situations. In such examples, the robot does not create an endearing perception to the medical workers, but has removed the nurses' agency to do the work that they do.

The third law highlights the time and effort needed to move robotics technologies from the research lab demonstrations and fancy videos to reliable products in the real world. Robots must work every time it is given a task. If it fails or if the results are unsatisfactory, the users will get frustrated and question whether the robot makes a person's life better or not. For robot technologies to mature, it would take a decade to achieve another 9 in reliability (i.e., from 99.9% to 99.99%).

17.4.3.2 The 3 laws and humanizing artificial touch

By aiming for humanlike appearance and touch, the book Humanizing Artificial Touch addresses the first law on the promise given by appearance. The visual appearance of an artificial hand should perform like a human hand. This book is an initial attempt to fulfill that promise through the techniques discussed, for example, on the construction of structures (i.e., 3D scanning to silicone molding workflow). The five fingers of the hand are supposed to grasp objects, perform most of the 33 grasp types in the taxonomy, which some of the state-of-the-art hands can do. More importantly, the book emphasizes on the fulfillment of the basic requirements of tactile sensing, which others may have missed: detection of contact of the fingertip and object, the object and the environment, slippage, and the local and global surface primitives.

For the second law on preserving people's agency, the book presented basic concepts, techniques, and case studies on how to achieve humanized artificial skin, tactile sensing, and robot hand structures. Bio-inspired humanoid robotic and prosthetic hands, which address the functional and social requirements, seek to ensure that the hands are helpful and reliable partners rather than as sources of frustration and embarrassment. Artificial hands that serve the users' needs as assistive devices or as collaborative partners can enhance human agency when the hands perform as responsive as they should in shared spaces and environments.

This book addresses Brooks' third law by providing detailed design and analytical methods for the interconnected artificial skin, sensing, and structural components. Some of those topics dealt with the properties of the soft skin materials for friction and impact absorption (e.g., compliance and conformance), for determining the optimal skin thickness to prevent sensor damage, for including flexible joints to prevent hyperextension that leads to damage of the fingers, for the life cycle calculation of the tendon cables, for the durability of the 3D printed structures, and a few other topics to support the hand's robustness. The book provides a framework for designing artificial hands that can meet stringent reliability requirements when operating in real-world environments.

17.5 Humanizing artificial touch (HAT) tests

Having gathered the most relevant works on imitation and prerequisite technologies to achieve advanced prosthetic and robotic hands for the future, we now fuse all those perspectives to propose two tests. The Humanizing Artificial Touch (HAT) Tests have static and dynamic touch tests, reflecting Mori's static and dynamic artifacts in the Uncanny Valley, Turing's two progressive tests with human-human and human-machine players, and Brooks' lab to the real-world requirements for reliability.

17.5.1 Static touch test

17.5.1.1 The test

The test is designed to determine if a human responder can distinguish between a human hand and an artificial hand (Fig. 17.2). The test involves: (A) the

Responder and (B) the Interrogator who touches the Responder with (a) the Interrogator's own hand, (b) a soft-cold hand, and (c) a soft-warm hand.

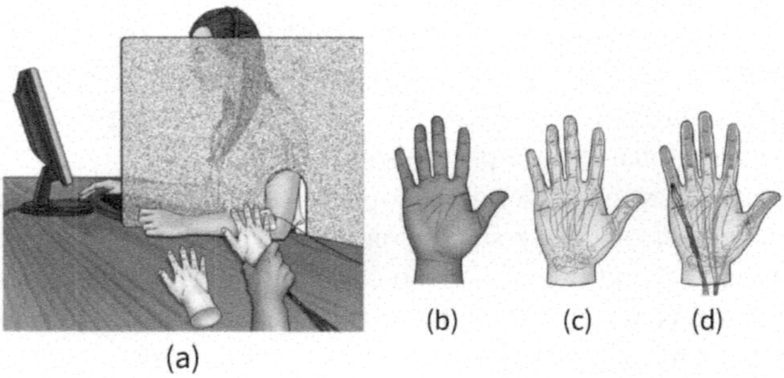

FIGURE 17.2 Static touch test. (a) Experimental setup where the test hands touched the participants. Various hands for the touch test: (b) human hand, (c) soft hand without heaters, and (d) soft and warm hand. Reproduced from Cabibihan et al. [10].

17.5.1.2 Setup

The Responder (A) is blindfolded or is seated in a way that prevents visual contact on the hands. Noise canceling headphones are used to remove any audible cues. The Interrogator (B), with his/her hand (a) and the two other artificial hands (b, c) touches the Responder's forearm with a slow, pressing motion. Some of the technical details are listed below:

- The ambient room temperature is set to a comfortable 21°C.
- Both artificial hands (b, c) are constructed with soft materials with Shore durometer values at the OO scale. The internal structure has a skeleton similar to human hands.
- The soft-cold hand (b) keeps the ambient temperature of the experiment room.
- The soft-warm hand (c) has embedded heaters with skin surface temperature of around 28°C.
- To make the touch consistent, a touch sensor can be mounted on the Responder's forearm. The Interrogator applies a contact force of the hands (a, b, c) to around 1 N as displayed in a display console.

17.5.1.3 Gameplay

The Responder (A) is tested with a series of touches with the three types of hands (a, b, c). The sequence is predetermined pseudorandomly. After each touch, the Interrogator (B) asks the Responder (A) to rate the touch based on the certainty of artificial or human touch. The 5-point Likert scale range from a score of 1 that implies that the touch was "very certain to be artificial" and 5 implying that the touch was "very certain to be human".

17.5.1.4 Goal

The goal is to make the Responder (A) believe that the touch from the soft-warm hand (c) is touch from the human hand (a). The main question to assess is whether the touch is human or not.

17.5.1.5 Evaluation

The success of the test is measured using Receiver Operating Characteristic (ROC) curve analysis. If the soft-warm artificial hand is indistinguishable from a human touch, the Area Under the Curve (AUC) for distinguishing the artificial hand from a human hand will be close to 0.5 or visually close to the diagonal line of the ROC curve.

17.5.1.6 Other variants

A variation of this test is a caress on the forearm, which is a form of a Sliding Touch Test.

- The Responder (A) and Interrogator (B) are those in relationships as a husband-wife pair or a girlfriend-boyfriend pair. The Interrogator will touch their partner-Responder in the forearm where the C-tactile (CT) afferents are known to be located. The CT afferents were reported to code pleasant tactile sensations [11]. The three hands (a, b, c) will be the same as in the Static Touch Test and will touch the Responder in the same starting location and contact force. The Interrogator will move the hands (a, b, or c) to a sliding motion at around 1-10 cm/s at the Responder's forearm.
- As another variation, artificial fingerprints can be added to the artificial hands (b, c). The human fingerprints and sweat contribute to the sliding contact on the skin [12]. The caressing touch with fingerprints could increase the realism and the pleasantness of the touch.
- The Responder's internal state can be tested to demonstrate if there is an emotional connection that can be observed from the human or artificial touch. Similarly, the likability or disgust that was described in the Uncanny Valley's hypothesis can be tested. Wearable sensors that detect the heart rate, galvanic skin response, and stress can be used to infer the affective state of the Respondent. Electroencephalography (EEG) can also be used to monitor the cognitive load, stress, and attention levels during the Static or Sliding Touch Test.

17.5.2 Dynamic touch test

17.5.2.1 The test

The test is designed to determine if a human responder can distinguish between the touch of a human hand and that of an artificial hand in a human-robot handshake test (Fig. 17.3). The test involves: (A) the Responder, (B) a Human handshaking partner, (C) a Robot handshaking partner, and (D) the Interrogator.

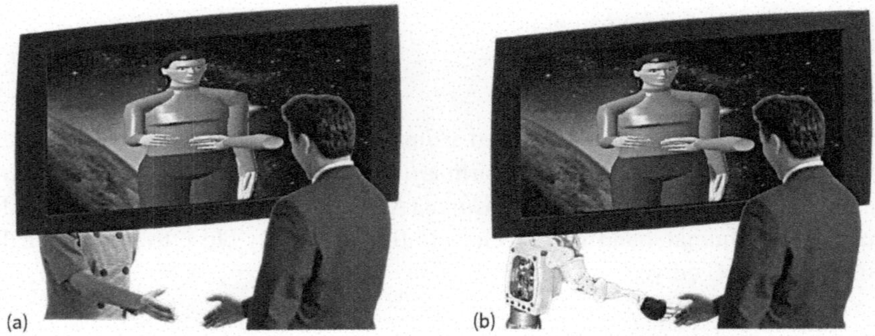

FIGURE 17.3 Dynamic touch test. An early conceptual depiction of a handshake touch test that isolates the perception of touch by using a Virtual Reality Cave Automatic Virtual Environment (VR-CAVE) in 2007. More recent virtual reality systems can use head-mounted consoles. (a) Human-human handshake. (b) Human-robot handshake.

17.5.2.2 Setup

To remove any visual biases, virtual reality system is used [13][14]. Preferably, the Responder wears a head-mounted display before entering the experiment room. Noise is likewise canceled with headphones to prevent any clues from the sound coming from the robot's motors. The Responder (A) approaches the test platform. The Interrogator (D) gives the instructions on whether the Human (B) or Robot (C) partner shakes hands first with the Responder (A). Some of the technical specifications are as follows [15] [16]:

- The conditions in the static test are repeated for the soft-warm hands.
- The robot hands need to be dexterous with finger flexion/extension and abduction/adduction, and wrist radial/ulnar deviation, flexion/extension, and pronation/supination. The intent is to fully and safely envelop the human partner's hand.
- The robot arm moves in three handshake phases: reaching, contact, and return. The full handshake is around 3.63 s (reaching 0.92 s, up and down motion 1.96 s, and return 0.75 s).
- Within the handshake motion, there is a 0.52 s of no motion during contact. This can be used for non-verbal communication, like looking eye-to-eye and smiling, to deepen emotional connection.
- The robot needs a vision system to predict the handshake partner's hand location to execute a reaching motion with human speed (e.g., 14 cm/s).
- The robot needs a control system, like high admittance control, to handle the variability in human handshakes. The robot will have to adapt its stiffness and response to different social cues during the handshake (e.g., vigorous, limp, neutral).

- Robots typically apply a uniform sinusoidal handshake motion. A human handshake has a gradual decay in amplitude. The peak starts at 5.3 cm and ends at 0.2 cm, with 2-4 oscillations.

17.5.2.3 Gameplay

The Responder (A) shakes hands with a Human (B) and a Robot (C) with a soft-warm hand. The handshake partner is predetermined pseudorandomly. After each handshake, the Interrogator (D) asks the Responder (A) to rate the touch on a 5-point Likert scale range from a score of 1 that implies that the handshake was "very certain to be artificial" and 5 implying that the handshake was "very certain to be human".

17.5.2.4 Goal

The goal is to make the Responder (A) believe that the handshake from the Robot (C) is a handshake from a Human (B). The main question to assess is whether the handshake is from a human or not.

17.5.2.5 Evaluation

Similar to the Static Touch Test, the success handshake is measured using Receiver Operating Characteristic (ROC) curve analysis. An indistinguishable robot handshake will have AUC 0.5 and the plotted curve is nearest to the ROC's diagonal line.

17.5.2.6 Other variants

There are a few suggested variants of the dynamic touch test that will be presented according to increasing difficulty.

- The easier variant could be the Fist Bump Test between the Responder (A) and the Human (B) and Robot (C) interaction partners. The fist bump explores the precise approach of the arm, the alignment of the fists, the safe contact force, and the rebound. This test can explore the safe fist bump forces that a Robot (C) applies to the Responder (A).
- The high-five, a ubiquitous gesture of celebration in sports and teamwork, involves a coordinated sequence of movements of the upper limb. In the High-Five Test, one interaction partner's hand must approach, locate, and orient the other partner's hand. The hand is positioned in a flat manner with the palm facing forward. The force applied by the Robot (C) during the high-five must be controlled so as not to hurt the Human (A) partner.
- Perhaps the most challenging handshake test is to replace the Robot with another Human (C) who is using a humanized artificial hand (i.e., an amputee). All the nuances of an advanced artificial hand design from skin mechanics, sensing, feedforward-feedback control, and brain-computer interface will be put to the test when an upper limb amputee

(C) directly shakes the hand of the Responder (A) in a VR environment similar to the standard handshake test. An ambitious goal is to trick the Responder (A) into believing that the handshake is from a realistic human hand and there are no significant differences with the handshakes of Human (B) and Human (C) with a humanized artificial hand. If this test is successfully passed, the limitations of artificial flesh (Turing), tactile sensing (Brooks), and prosthetic hand (Mori) would have been addressed.

17.6 Epilogue

In 1962, an archaeological team led by Daniel Oates made a remarkable discovery in an excavation in Iraq [17]. That finding is now considered to be one of the oldest records of the handshake. The artifact, a bas-relief sculpture, dates back to the 9th century BC in the ancient Assyrian city of Nimrud. The sculpture depicts two kings and their followers. Marduk-zakir-sumi I, the Babylonian King, and Shalmanaser III, the Assyrian King, appear to be in a handclasp to signify an alliance between their two kingdoms (Fig. 17.4).

FIGURE 17.4 The first recorded handshake in human history. King Marduk-zakir-sumi I of Babylon (855–819 BC; center-left) and King Shalmanaser III of Assyria (859–824 BC; center-right) in a hand clasp. In the 1962 excavations in present-day Iraq, the bas relief was discovered in a raised platform in front of a ceremonial hall in Nimrud, a city in ancient Assyria. Nearly 3,000 years later, that handclasp evolved into the modern handshake. Reproduced from Oates [17], with permission.

From Mesopotamia, the practice of the handclasp traveled west. In ancient Greece, the hand gesture became known as *dexiosis*, which is the action of two people offering their right hands to one another [18]. This act was known to symbolize trust, peace, and equality. *Dexiosis* has been depicted in Greek art from the 5th century BC.

The ancient Romans later adopted and codified the gesture around 2nd century BC. *Dextrarum iunctio*, which is Latin for joining of the right hands, became a ceremonial part of Roman marriage rites [19]. Because the right hand was considered to be sacred to the deity Fides, the joining of hands by the wedded couple symbolized a divine pledge of loyalty [20]. In the 1800s, the handclasp became an alternative to the bowing or tipping of the hat during the Victorian era [21]. The etiquette manuals even included the proper technique. The Victorian shake must be firm, yet not overly strong.

The ancient handclasp, *dexiosis*, *dextrarum iunctio*, and the Victorian shake have all evolved into the modern handshake. This gesture is being exchanged when meeting or parting with someone, communicating appreciation, sympathy, and gratitude. A firm handshake can communicate sociability, competence, and trustworthiness while a poor one can communicate introversion and shyness [22][23].

In the 1950s, Alan Turing famously asked: "**Can machines think?**". By 2025, advanced Artificial Intelligence has demonstrated that it can. AI has blurred the lines of this once-hypothetical question. Turing also stated that no engineer or chemist during his time could produce a material that is indistinguishable from the human skin [2]. He noted that there was no point in trying to make a "thinking machine" more human by dressing it up in artificial flesh. In 2015, we showed that the touch from an artificial soft and warm skin can be indistinguishable from a human touch [24].

The fabric of society has evolved. The handclasp of ancient kings, once a gesture of statecraft, transformed into the *dexiosis* of the Greeks, the *dextrarum iuntio* of the Romans and the Victorian shake in England. Now, many forms of social touch are exchanged: the modern handshakes for greetings, the affectionate hugs for consolation and love, the high fives for celebrations, the pat in the back for a job well-done, and the fist bumps during the 2020 pandemic.

We are seeing an ever-pressing hunger for human connection; there is a biological and physical need for the human touch. That need will inevitably influence the design of artificial hands for the war-wounded, trauma patients, for those with congenital differences, and for the hands of social and humanoid robots.

We are entering the next era of humanized artificial touch. This era will attempt to close the loop: from learning about the human flesh, receptors, and bones to crafting artificial skin, sensors, and structures. It leads us back to asking ourselves whether the artificial is indistinguishable from the real: "**Is the touch human or not**?" When combined with the thinking machines running on Artificial Intelligence and embodied by Robots, the next question evolves into something more profound or perhaps more disturbing for others: "**Is it human (like) or not?**".

References

[1] R.S. Johansson, Sensory Input and Control of Grip, Wiley, 2007, pp. 45–63.

[2] A.M. Turing, Computing Machinery and Intelligence, Mind LIX (1950) 433–460, https://doi.org/10.1093/mind/lix.236.433.

[3] C.R. Jones, I. Rathi, S. Taylor, B.K. Bergen, People Cannot Distinguish GPT-4 from a Human in a Turing Test, Proceedings of the 2025 ACM Conference on Fairness, Accountability, and Transparency. FAccT'25, ACM, 2025, pp. Jun:1615–39.

[4] M. Mori, The Uncanny Valley, vol. 7, Energy, 1970, pp. 33–35.

[5] M. Mori, K. MacDorman, N. Kageki, The uncanny valley [from the field], IEEE Robotics & Automation Magazine 19 (2012 Jun) 98–100, https://doi.org/10.1109/mra.2012.2192811.

[6] J.J. Cabibihan, M.C. Carrozza, P. Dario, S. Pattofatto, M. Jomaa, A. Benallal, The Uncanny Valley and the Search for Human Skin-Like Materials for a Prosthetic Fingertip, in: 2006 6th IEEE-RAS International Conference on Humanoid Robots, 2006, pp. 474–477.

[7] E. Poliakoff, N. Beach, R. Best, T. Howard, E. Gowen, Can looking at a hand make your skin crawl? Peering into the uncanny valley for hands, Perception 42 (2013 Jan) 998–1000, https://doi.org/10.1068/p7569.

[8] E. Poliakoff, S. O'Kane, O. Carefoot, P. Kyberd, E. Gowen, Investigating the uncanny valley for prosthetic hands, Prosthetics and Orthotics International 42 (2018 Feb) 21–27, https://doi.org/10.1177/0309364617744083.

[9] R. Brooks, Rodney Brooks's Three Laws of Robotics: What to Know About Real Robots Deployed in the Real World. 2024 Aug.

[10] J.J. Cabibihan, D. Joshi, Y.M. Srinivasa, M.A. Chan, A. Muruganantham, Illusory sense of human touch from a warm and soft artificial hand, IEEE Transactions on Neural Systems and Rehabilitation Engineering 23 (2015) 517–527, https://doi.org/10.1109/tnsre.2014.2360533.

[11] L.S. Löken, J. Wessberg, I. Morrison, F. McGlone, H. Olausson, Coding of pleasant touch by unmyelinated afferents in humans, Nature Neuroscience 12 (2009 Apr) 547–548, https://doi.org/10.1038/nn.2312.

[12] D. Kim, D. Yun, A study on the effect of fingerprints in a wet system, Scientific Reports 9 (2019 Nov), https://doi.org/10.1038/s41598-019-51694-9.

[13] A. Gaballa, R.S. Cavalcante, E. Lamounier, A. Soares, J.J. Cabibihan, Extended reality "X-reality" for prosthesis training of upper-limb amputees: a review on current and future clinical potential, IEEE Transactions on Neural Systems and Rehabilitation Engineering 30 (2022) 1652–1663, https://doi.org/10.1109/tnsre.2022.3179327.

[14] N. Mazhar, A. Gaballa, A.K. Pandey, J.J. Cabibihan, Virtual Reality Hand Tracking for Immersive Telepresence in Rehabilitative Serious Gaming, Social Robotics, Springer Nature, Singapore, 2023, pp. 3–12, https://doi.org/10.1007/978-981-99-8715-3_1.

[15] J.J. Cabibihan, R. Pradipta, Y.Z. Chew, S.S. Ge, Towards humanlike social touch for prosthetics and sociable robotics: handshake experiments and finger phalange indentations, in: Advances in Robotics, Springer, Berlin, Heidelberg, 2009, pp. 73–79, https://doi.org/10.1007/978-3-642-03983-6_11.

[16] J.J. Cabibihan, A. El-Noamany, A.M.R.M. Ahmed, M.H. Ang, Guidelines for robot-to-human handshake from the movement nuances in human-to-human handshake, Frontiers in Robotics and AI 9 (2022), https://doi.org/10.3389/frobt.2022.758519.

[17] D. Oates, The excavations at Nimrud (Kalhu), 1962, Iraq 25 (1963) 6–37, https://doi.org/10.2307/4199729.

[18] L. Novakova, M. Pagacova, Dexiosis a meaningful gesture of the classical antiquity, ILIRIA International Review 6 (2016 Jul) 209, https://doi.org/10.21113/iir.v6i1.213.

[19] K.K. Hersch, The Roman Wedding: Ritual and Meaning in Antiquity, Cambridge University Press, Cambridge, England, 2010 May.

[20] S.D. Ricks, Dexiosis and dextrarum iunctio: the sacred handclasp in the classical and early christian world, The FARMS Review 18 (2006) (2006 Jan) 431–435, https://doi.org/10.5406/farmsreview.18.1.0431.

[21] T. Lundmark, Getting the shakes: the story of the handshake, in: Tales of Hi and Bye: Greeting and Parting Rituals Around the World, Cambridge University Press, 2009, pp. 54–70, https://doi.org/10.1017/CBO9780511657450.008.
[22] J. Astrom, L.H. Thorell, Greeting behaviour and psychogenic need: interviews on experiences of therapists, clergymen, and car salesmen, Perceptual and Motor Skills 83 (1996) 939–956, https://doi.org/10.2466/pms.1996.83.3.939.
[23] W.F. Chaplin, J.B. Phillips, J.D. Brown, N.R. Clanton, J.L. Stein, Handshaking gender, personality, and first impressions, Journal of Personality and Social Psychology 79 (2000) 110–117, https://doi.org/10.1037/0022-3514.79.1.110.
[24] J.J. Cabibihan, S.S. Chauhan, Physiological responses to affective tele-touch during induced emotional stimuli, IEEE Transactions on Affective Computing 8 (2017) 108–118, https://doi.org/10.1109/taffc.2015.2509985.

Index

G

H

I

Printed and bound by CPI Group (UK) Ltd, Croydon, CR0 4YY
17/09/2026
02218538-0001